APPROXIMATION
OF FUNCTIONS

APPROXIMATION

OF FUNCTIONS

BY

G. G. LORENTZ

Professor of Mathematics
University of Texas
Austin

CHELSEA PUBLISHING COMPANY

NEW YORK, N. Y.

SECOND EDITION

Copyright ©, 1966, by G. G. Lorentz
Copyright ©, 1986, by G. G. Lorentz

Library of Congress Catalog Card No. 85- 72465
International Standard Book No. 0-8284-0322-8

Printed on 'long-life' acid-free paper

Printed in the United States of America

Preface to the Second Edition

My book, Approximation of Functions (Holt, Rinehart and Winston), has been out of print for a number of years. But a considerable demand for it still remains which, I hope, is in some measure justified. A Chinese translation of the book has appeared recently.

I have therefore been happy to accept an offer by Chelsea Publishing Company to publish a second edition of the book. The changes are minimal. Some errors have been corrected and a faulty result about rational approximation has been replaced by a beautiful theorem of my friend, the late Geza Freud. I am grateful to D. D. Stancu, Cluj, Rumania for many corrections.

G. G. Lorentz

Preface to the First Edition

My purpose has been to write an easily accessible book on the approximation of functions that is simple and without unnecessary details, and is also complete enough to include the main results of the theory, including some recent ones. In some cases (for example, Chapter 7, saturation classes), this has been made possible by restricting discussions to a few representative theorems from a field. The leitmotiv of the book is that of the degree of approximation. Only Chapter 2 (Chebyshev's theorem and related results), Chapter 3 (auxiliary results and notions), and Chapter 11 do not depend on this idea. The justification of the latter chapter lies in its coverage of some applications of entropy, which are significant because of Kolmogorov's theorem.

Except for a few sections, only functions of a real variable have been treated. The beautiful results of Runge, Bernstein, Walsh, Mergeljan, Dzjadyk, and others in the complex domain remain outside the scope of this book.

The book can be used as a textbook for a graduate or an advanced undergraduate course, or for self-study. Notes at the end of each chapter give information about important topics not treated in the main text. Problems serve as illustrations; some of them are not easy. It was felt that it is more useful to solve one difficult problem than several easy ones. The Bibliography is not all-inclusive. It has been limited to works that can be expected to be particularly useful for the reader, and to others of utmost historical interest.

I owe thanks to my friends, colleagues, and students, who have helped me with my work. Above all, I must thank Professor E. Hewitt, the editor of this series, Professors G. T. Cargo, G. F. Clements, H. S. Shapiro, Messrs. J. Case and J. T. Scheick, all students in my class on approximation theory, and the OSR of the U.S. Air Force, whose grant supported my work. I would be grateful for any suggestions that readers may send to me.

G. G. Lorentz

Contents

[1]

Possibility of Approximation

1. Basic Notions

The problem of linear approximation can be described in the following way: Let Φ be a set of functions, defined on a fixed space A. If a function f on A is given, can one find a linear combination $P = a_1\phi_1 + \cdots + a_n\phi_n$, $\phi_i \in \Phi$, which is close to the function f? Two preliminary problems arise: We must select the set Φ, and also decide how the deviation of P from f should be measured.

We begin with the second question. Let A be a compact Hausdorff topological space,[1] and let $C = C[A]$ be the set of all continuous real functions on A. The set C is a linear space over the reals: sums $f + g$ and products af with real a and $f, g \in C$ belong to C and satisfy the axioms of a linear space. The supremum

$$\|f\| = \sup_{x \in A} |f(x)| \tag{1}$$

is attained for all functions $f \in C$; thus, $\|f\| = \max_{x \in A} |f(x)|$. This supremum has the following properties, which define a norm on $C[A]$:

$$\|f\| \geqslant 0; \qquad \|f\| = 0, \quad \text{if and only if} \quad f = 0; \tag{2}$$

$$\|af\| = |a| : \|f\|; \tag{3}$$

$$\|f + g\| \leqslant \|f\| + \|g\|. \tag{4}$$

Thus, C is a *normed linear space*. Similarly, the space of all continuous complex functions f on A with norm (1), which is also denoted by $C[A]$, is a normed linear space over the complex number field.

Unless something to the contrary is said, all our functions and scalars will be real. The convergence $f_n \to f$ in the norm of C, that is, $\|f_n - f\| \to 0$ as $n \to \infty$, is equivalent to the *uniform convergence* of $f_n(x)$ to $f(x)$ for all $x \in A$. It follows from this interpretation that the space C is complete: If f_n is a Cauchy sequence (that is, $\|f_n - f_m\| \to 0$ for $n, m \to \infty$), then f_n converges to some element f of C:

$$\|f_n - f\| \to 0. \tag{5}$$

---

[1] Without essential loss, the reader can substitute for this, here and in the remainder of the book, a compact metric space, or even a compact subset of a euclidean space.

Complete normed linear spaces are called *Banach spaces*. Many types of Banach spaces are important in the theory of approximation; for example, the spaces $L^p = L^p[a, b]$, $p \geqslant 1$, with the norm

$$\| f \| = \left\{ \int_a^b | f(x) |^p \, dx \right\}^{1/p}.$$

However, approximation in the spaces C remains both the most interesting and the most important special case (if one excludes the theory of orthogonal polynomials in the space L^2), and this book is devoted almost entirely to it.

The following definitions apply to any Banach space X with elements f and a distinguished subset Φ. We call f *approximable* by linear combinations

$$P = a_1\phi_1 + a_2\phi_2 + \cdots + a_n\phi_n, \qquad \phi_i \in \Phi, \qquad a_i \text{ real}, \tag{6}$$

if for each $\epsilon > 0$ there is a P with $\| f - P \| < \epsilon$. Often, Φ is a sequence: $\phi_1, \phi_2, \cdots, \phi_n, \cdots$. Then

$$E_n(f) = E_n^\Phi(f) = \inf_{a_1, \ldots, a_n} \| f - (a_1\phi_1 + \cdots + a_n\phi_n) \| \tag{7}$$

is the *n*th *degree of approximation* of f by the ϕ_i. If the infimum in (7) is attained for some P, this P is called *a linear combination of best approximation*. There is an exception to this notation: If the P are algebraic or trigonometric polynomials of a given degree, then n in (7) will refer to the degree of the polynomials rather than to the number of functions ϕ_i.

For the space $C[a, b]$ of continuous real functions on $[a, b]$, a natural sequence Φ is given by the powers $1, x, \cdots, x^n, \cdots$. In this case, the linear combinations of the first $n + 1$ functions are the *algebraic polynomials of degree n*: $P_n(x) = a_0 + a_1x + \cdots + a_nx^n$. In this definition we do *not* require that $a_n \neq 0$. A similar remark applies to some later definitions.

Another important compact set K is the additive group of real numbers $R = (-\infty, +\infty)$, taken modulo 2π; the distance $| x - x' |$ between $x, x' \in K$ is the minimal distance between the representations of x, x' in R. For obvious reasons, we can call K the *unit circle*. This K is a metric space with the distance $| x - x' |$. We shall follow the practice of identifying functions $f \in C^* = C[K]$ with the continuous 2π-periodic functions on R. For such functions, $\int_K f dx$ is the integral of f over any interval of R of length 2π. A function $f \in C^*$ does not necessarily have an indefinite integral F in C^*, for $\int_0^x f(t) \, dt$ is not necessarily periodic. Clearly, an $F \in C^*$ exists if and only if f has *mean-value zero*; that is, if $(1/2\pi) \int_K f dt = 0$. In this case, $F(x) = \text{const} + \int_0^x f dt$. Among these F there is *exactly one* with mean-value zero. Iterating this, we see that for each $p = 1, 2, \cdots$, a function $f \in C^*$ with mean-value zero has a family of *p*th indefinite integrals, which depends upon one additive constant. We shall call each of these integrals *a pth indefinite integral of f*.

A tool of approximation for functions $f \in C^*$ is the following set of *trigonometric polynomials*:

$$T_n(x) = \frac{a_0}{2} + a_1 \cos x + b_1 \sin x + \cdots + a_n \cos nx + b_n \sin nx. \qquad (8)$$

The polynomial (8) is said to have *degree n*. It is *even* (or *odd*) if only cosines $\cos kx$, $k = 0, \cdots, n$ (or sines) appear in the representation. Simple trigonometric formulas imply that the product of two trigonometric polynomials of degrees n and m is equal to a trigonometric polynomial of degree $n + m$.

In analogy with these two cases, the linear combinations (6) will also sometimes be called polynomials.

In the present chapter, we shall discuss the possibility of approximation of functions by polynomials. In Chapter 2, properties of polynomials of best approximation will be treated.

2. Linear Operators

The following theorem (which will be proved in Sec. 3) is due to Weierstrass [103]:

THEOREM 1.   Each continuous real function f on $[a, b]$ is approximable by algebraic polynomials: For each $\epsilon > 0$ there is a polynomial $P_n(x) = \sum_0^n a_k x^k$ with

$$|f(x) - P_n(x)| < \epsilon, \qquad a \leqslant x \leqslant b. \qquad (1)$$

The most natural way to prove a theorem of this type is to give an explicit formula for the polynomial $P_n(x)$. In terms of f, this formula is usually linear.

A function $g = L(f)$ from a Banach space X into a Banach space Y is called a *linear operator* if it satisfies the conditions

$$L(f + f') = L(f) + L(f'); \qquad L(af) = aL(f) \qquad (2)$$

for all $f, f' \in X$ and all real a. If Y is the real line R, L is called a *linear functional*. A linear operator L is called *bounded* if

$$\|L(f)\| \leqslant M \|f\|, \qquad f \in X \qquad (3)$$

for some positive constant M. In this case, the infimum of all M for which (3) is true is still an admissible M. This minimal M is called the *norm* of L, and is denoted by $\|L\|$. From this definition it follows that

$$\|L\| = \sup_{f \neq 0} \frac{\|L(f)\|}{\|f\|} = \sup_{f \neq 0} \left\| L\left(\frac{f}{\|f\|}\right) \right\| = \sup_{\|f\|=1} \|L(f)\|. \qquad (4)$$

A bounded linear operator is continuous: From $f_n \to f$ (in the norm of X), it follows that $L(f_n) \to L(f)$ (in the norm of Y), since

$$\|L(f_n) - L(f)\| \leqslant \|L\| \cdot \|f_n - f\|.$$

In the case $X = C[A]$, we can distinguish *positive* elements of C: We write $f \geqslant 0$ if $f \in C$ and $f(x) \geqslant 0$ for all $x \in A$. An operator L that maps C into itself is called a *positive operator* if it transforms each positive element f into a positive element g. For a positive linear operator, we have $L(f) \leqslant L(g)$ if $f \leqslant g$ (that is, if $g - f \geqslant 0$); also, $|L(f)| \leqslant L(|f|)$, where $|f|$ is the function with values $|f(x)|$. An operator of this type is always bounded: From

$$|L(f)| \leqslant L(|f|) \leqslant L(\|f\| e) = \|f\| L(e)$$

(here e is the constant function $e(x) = 1$), it follows that

$$\|L(f)\| \leqslant \|L(e)\| \cdot \|f\|,$$

so that

$$\|L\| = \|L(e)\|.$$

For the value of $L(f)$ at $x \in A$, we write $L(f, x)$.

EXAMPLES

1. Bernstein Polynomials. For a function f defined on $[0, 1]$, let

$$B_n(f, x) = \sum_{k=0}^{n} \binom{n}{k} x^k (1 - x)^{n-k} f\left(\frac{k}{n}\right), \qquad n = 0, 1, \cdots. \tag{5}$$

Clearly, this is a positive linear operator which maps $C[0, 1]$ into itself. By the binomial formula,

$$B_n(e, x) = \sum_{k=0}^{n} p_{nk}(x) = 1, \qquad p_{nk}(x) = \binom{n}{k} x^k (1 - x)^{n-k}, \tag{6}$$

and hence $\|B_n\| = 1$, for $n = 0, 1, \cdots$.

2. Fourier Series. Let f be a 2π-periodic, integrable function. The coefficients of its *Fourier series*,

$$\frac{a_0}{2} + \sum_{k=1}^{\infty} (a_k \cos kx + b_k \sin kx), \tag{7}$$

are given by the formulas

$$a_k = \frac{1}{\pi} \int_{-\pi}^{\pi} f(t) \cos kt \, dt, \qquad b_k = \frac{1}{\pi} \int_{-\pi}^{\pi} f(t) \sin kt \, dt. \tag{8}$$

For example, the finite sum 1(8), augmented by zero terms, is the Fourier series of the trigonometric polynomial T_n. We consider the nth partial sum s_n of the series (7). A standard computation gives

$$s_n = s_n(f, x) = \frac{1}{\pi} \int_{-\pi}^{\pi} f(t) \frac{\sin (2n + 1) \dfrac{t - x}{2}}{2 \sin \dfrac{t - x}{2}} \, dt. \tag{9}$$

To obtain this, one writes $s_n = u_0 + u_1 + \cdots + u_n$, where

$$u_0 = \frac{1}{\pi} \int_{-\pi}^{\pi} \tfrac{1}{2} f(t) \, dt, \qquad u_k = \frac{1}{\pi} \int_{-\pi}^{\pi} f(t) \cos k(t - x) \, dt, \quad k = 1, 2, \ldots ,$$

and applies the formula

$$\frac{1}{2} + \cos \alpha + \cdots + \cos n\alpha = \frac{\sin (2n + 1) \, (\alpha/2)}{2 \sin (\alpha/2)} = D_n(\alpha), \qquad (10)$$

which is obtained by multiplying both sides with $2 \sin (\alpha/2)$. In the same way, by means of the formula

$$\sin \frac{\alpha}{2} + \sin \frac{3}{2} \alpha + \cdots + \sin (2n - 1) \frac{\alpha}{2} = \frac{\sin^2 (n\alpha/2)}{\sin (\alpha/2)} , \qquad (11)$$

we obtain a representation of the arithmetic mean σ_n of the s_n:

$$\sigma_n = \sigma_n(f, x) = \frac{s_0 + \cdots + s_{n-1}}{n} = \frac{1}{2\pi n} \int_{-\pi}^{\pi} f(t) \left(\frac{\sin \dfrac{n(t - x)}{2}}{\sin \dfrac{t - x}{2}} \right)^2 dt. \qquad (12)$$

Thus, $\sigma_n(f)$ is a sequence of positive linear operators, mapping C^* into itself. We have $\| \sigma_n \| = 1$, since $\sigma_n(e) = e$ (for the function $e(x) \equiv 1$, all $s_n(x) \equiv 1$). The operators $s_n(f), f \in C^*$ are also linear, but not positive. It follows from the definition of s_n and σ_n that for each f, both $s_n(f, x)$ and $\sigma_n(f, x)$ are trigonometric polynomials of degrees n and $n - 1$, respectively. The norm of $s_n(f)$ is given by the following theorem.[2]

THEOREM 2 (Fejér). The norm $\| s_n \|$ of the operator $s_n(f)$ is equal to

$$A_n = \frac{1}{\pi} \int_{-\pi}^{\pi} \left| \frac{\sin (2n + 1) \, (t/2)}{2 \sin (t/2)} \right| dt = \frac{4}{\pi^2} \log n + O(1); \qquad (13)$$

also, the norm of $s_n(f, x)$, for each fixed x, considered as a linear functional from C^* to R, is equal to (13).

Proof. Since $D_n(t)$ has period 2π,

$$| s_n(f, x) | \leqslant \| f \| \frac{1}{\pi} \int_{-\pi}^{\pi} | D_n(t - x) | \, dt = A_n \| f \| .$$

---

[2] We use the following notation: If u_n and $v_n > 0$ are functions of n, we write (a) $u_n = O(v_n)$ if $| u_n | \leqslant M v_n$ for some constant M; (b) $u_n = o(v_n)$ if $u_n/v_n \to 0$ as $n \to \infty$; (c) $u_n \sim v_n$ if $u_n/v_n \to 1$ as $n \to \infty$, and (d) $u_n \approx v_n$ if u_n/v_n is contained between two constants m, M, where $0 < m < M$.

This inequality shows that the functionals as well as the operator s_n have norms not exceeding A_n. To show that these norms are actually equal to A_n, it is sufficient to find, for given $x \in K$ and $\epsilon > 0$, a continuous function g for which $\| g \| = 1$ and

$$s_n(g, x) = \frac{1}{\pi} \int_{-\pi}^{\pi} g(t) \, D_n(t - x) \, dt > A_n - \epsilon. \tag{14}$$

For the function $g_0(t) = \operatorname{sign} D_n(t - x)$, we have

$$\frac{1}{\pi} \int_{-\pi}^{\pi} g_0(t) \, D_n(t - x) \, dt = A_n \, .$$

But g_0 is not continuous; it has jump discontinuities at the finitely many points t_ν of $[-\pi, \pi]$, where $D_n(t - x)$ changes sign. We surround each t_ν by a small interval $I_\nu = (t_\nu - \delta, t_\nu + \delta)$ and change g_0 on each I_ν so as to obtain a continuous function g, which has values between -1 and $+1$ everywhere and coincides with g_0 outside the I_ν. The difference between the integrals $\int_{-\pi}^{\pi} g D_n dt$ and $\int_{-\pi}^{\pi} g_0 D_n dt$ does not exceed $2 \int_E | D_n(x - t) | \, dt$, where E is the union of the intervals I_ν. If δ is sufficiently small, we have (14).

To obtain an asymptotic formula for A_n, we write $A_n = \pi^{-1} \int_0^\pi 2 \, | D_n(t) | \, dt$, since $D_n(t)$ is even. The function under the integral sign is equal to

$$\left| \cot \frac{t}{2} \sin nt + \cos nt \right| = \left| \frac{2}{t} \sin nt + \left(\cot \frac{t}{2} - \frac{2}{t} \right) \sin nt + \cos nt \right|.$$

Since $\cot u - u^{-1}$ is bounded in $(0, \pi/2)$, we have

$$A_n = \frac{2}{\pi} \int_0^\pi \frac{| \sin nt |}{t} \, dt + O(1).$$

The integral of $t^{-1} | \sin nt |$ over $(0, \pi/n)$ is bounded, since $| \sin nt | \leqslant nt$. Thus,

$$A_n = \frac{2}{\pi} \sum_{k=1}^{n-1} \int_{k\pi/n}^{(k+1)\pi/n} \frac{| \sin nt |}{t} \, dt + O(1)$$

$$= \frac{2}{\pi} \int_0^{\pi/n} \sin nt \sum_{k=1}^{n-1} \frac{1}{t + n^{-1} k\pi} \, dt + O(1).$$

Let $S(t)$ denote the last sum. For $0 \leqslant t \leqslant \pi/n$, $S(t)$ lies between

$$S(0) = n\pi^{-1} \left(1 + \frac{1}{2} + \cdots + \frac{1}{n-1} \right) \quad \text{and} \quad S(\pi/n) = S(0) + O(n).$$

If we use the facts that

$$1 + \frac{1}{2} + \cdots + \frac{1}{n-1} = \log n + O(1),$$

and that

$$\int_0^{\pi/n} \sin ntdt = \frac{2}{n},$$

we obtain:

$$A_n = \frac{4}{\pi^2} \log n + O(1). \qquad\qquad \text{| (15)}$$

3. Approximation Theorems

The operator s_n is not suitable for the uniform approximation of arbitrary continuous functions, as we shall see in Chapter 6. However, both B_n and σ_n can be used for this purpose.

It has been observed by Bohman and Korovkin [7] that for a sequence L_n of positive linear operators, convergence often can be established quite simply by checking it for certain finite sets of functions f. Let A be a compact Hausdorff topological space (with at least two points). Let $f_1, \cdots, f_m$ be continuous real functions on A that have the following property[3]:

there exist continuous real functions $a_i(y), y \in A, i = 1, \cdots, m$ such that

$$P_y(x) = \sum_{i=1}^m a_i(y) f_i(x) \qquad\qquad (1)$$

is positive, and equal to zero if and only if $x = y$.

THEOREM 3. If the functions $f_1, \cdots, f_m$ satisfy (1) and if L_n is a sequence of positive linear operators that map $C[A]$ into itself and satisfy

$$L_n(f_i, x) \to f_i(x) \quad \text{uniformly for} \quad x \in A, \qquad i = 1, \cdots, m, \qquad (2)$$

then

$$L_n(f, x) \to f(x) \quad \text{uniformly in } x \text{ for each} \quad f \in C[A]. \qquad (3)$$

Proof. We begin with some properties of the functions $P(x) = \sum_1^m a_i f_i(x)$. There exists a $\bar{P}$ with $\bar{P}(x) > 0$ for all $x \in A$: if $y_1 \neq y_2$ are two points of A, we can take $\bar{P} = P_{y_1} + P_{y_2}$. From (2) we have $L_n(P, x) \to P(x)$ uniformly in x for each P with constant coefficients. We also have

$$L_n(P_y, y) = \sum_{i=1}^m a_i(y) L_n(f_i, y) \to \sum_{i=1}^m a_i(y) f_i(y) = 0,$$

---

[3] The assumption that $a_i(y)$ are continuous could be omitted, but it simplifies the proof of Theorem 3.

and the convergence is uniform in y because the $a_i(y)$ are bounded. Finally, for some constant $M_0 > 0$, $\| L_n(e) \| \leqslant M_0$. This follows from

$$L_n(e, x) \leqslant a L_n(\bar{P}, x) \to a\bar{P}(x),$$

where $a > 0$ is taken so that $1 = e(x) \leqslant a\bar{P}(x)$, $x \in A$.

LEMMA.   Let $f_y \in C[A]$, $y \in A$, be a family of functions for which $f_y(x)$ is a continuous function of the point $(x, y) \in A \times A$ and $f_y(y) = 0$ for all $y \in A$. Then

$$L_n(f_y, y) \to 0 \qquad \text{uniformly in } y. \tag{4}$$

Proof.   Consider the "diagonal" set $B = \{(y, y)\}$ in $A \times A$ and some $\epsilon > 0$. Each point of B has a neighborhood U in $A \times A$ for which $|f_y(x)| < \epsilon$ if $(x, y) \in U$. The union G of all these U is an open set; its complement F is compact. Let

$$m = \min_{(x,y)\in F} P_y(x) > 0, \qquad M = \max_{(x,y)\in F} |f_y(x)|.$$

For all x, y we have

$$|f_y(x)| < \epsilon + \frac{M}{m} P_y(x). \tag{5}$$

In fact, $|f_y(x)|$ does not exceed the first term on the right if $(x, y) \in G$, nor the second term if $(x, y) \in F$. From (5) we derive

$$|L_n(f_y, y)| \leqslant \epsilon L_n(e, y) + \frac{M}{m} L_n(P_y, y)$$

$$\leqslant M_0\epsilon + \frac{M}{m} L_n(P_y, y) \leqslant (M_0 + 1)\,\epsilon,$$

for all large n.   ∎

Now the proof of the theorem can be completed easily. If $f \in C[A]$ is given, we put

$$f_y(x) = f(x) - \frac{f(y)}{\bar{P}(y)}\,\bar{P}(x).$$

By the lemma,

$$L_n(f, y) - \frac{f(y)}{\bar{P}(y)} L_n(\bar{P}, y) \to 0,$$

and since $L_n(\bar{P}, y) \to \bar{P}(y)$, we obtain (3).   ∎

For example, if $A = [a, b]$, the system $f_1 = 1, f_2 = x, f_3 = x^2$ satisfies the condition (1). We can take

$$P_y(x) = (y - x)^2 = y^2 f_1 - 2y f_2 + f_3.$$

Theorem 3 allows us to check the convergence of certain operators with a minimum of computations. We begin by proving Theorem 1. The linear substitution $t = (x - a)/(b - a)$ reduces the interval $a \leqslant x \leqslant b$ to the interval $0 \leqslant t \leqslant 1$. Thus, Theorem 1 follows from

THEOREM 4. If the function f is continuous on $[0, 1]$, then

$$\lim_{n \to \infty} B_n(f, x) = f(x) \qquad \text{uniformly for} \qquad 0 \leqslant x \leqslant 1. \tag{6}$$

Proof. For the polynomials p_{nk} of 2(6), we have

$$\sum_{k=0}^{n} k p_{nk}(x) = \sum_{k=1}^{n} k \binom{n}{k} x^k (1 - x)^{n-k}$$

$$= nx \sum_{l=0}^{n-1} \binom{n-1}{l} x^l (1 - x)^{n-1-l} = nx; \tag{7}$$

$$\sum_{k=0}^{n} k(k - 1) p_{nk}(x) = n(n - 1) x^2 \sum_{l=0}^{n-2} \binom{n-2}{l} x^l (1 - x)^{n-2-l} = n(n - 1) x^2,$$

so that

$$\sum_{k=0}^{n} k^2 p_{nk}(x) = n^2 x^2 + nx(1 - x). \tag{8}$$

Formulas 2(6), (7), and (8) mean that the functions $f_1 = 1$, $f_2 = x$, and $f_3 = x^2$ have as their Bernstein polynomials, respectively, 1, x, and $x^2 + n^{-1}x(1 - x)$, for $n \geqslant 2$. Conditions (1) and (2) hold, and (6) follows from Theorem 3. ∎

Useful in connection with Bernstein polynomials are the sums

$$T_{nr}(x) = \sum_{k=0}^{n} (k - nx)^r p_{nk}(x), \qquad r = 0, 1, \cdots. \tag{9}$$

Using formulas (7) and (8), we see that $T_{n0} = 1$, $T_{n1} = 0$, $T_{n2} = nx(1 - x)$. In order to compute the T_{nr} for $r \geqslant 2$, it is convenient to use the recurrence relation

$$T_{n,r+1} = x(1 - x) (T'_{nr} + nr T_{n,r-1}), \qquad r \geqslant 1, \tag{10}$$

which follows from (9) by differentiation, if one notices that

$$\frac{d}{dx} p_{nk}(x) = \frac{k - nx}{x(1 - x)} p_{nk}(x). \tag{11}$$

From (9) we obtain

$$T_{n0} = 1, \qquad T_{n1} = 0, \qquad T_{n2} = nX, \qquad T_{n3} = n(1 - 2x) X,$$
$$T_{n4} = 3n^2 X^2 - 2nX^2 + nX(1 - 2x)^2, \qquad X = x(1 - x). \tag{12}$$

It follows, for instance, that $T_{n4}(x) \leqslant Mn^2$, $0 \leqslant x \leqslant 1$. We can apply this in order to estimate, for a fixed $\delta > 0$, the sum of terms $p_{nk}(x)$ extended over the values of k for which $| x - k/n | \geqslant \delta$:

$$\sum_{|x-k/n| \geqslant \delta} p_{nk}(x) \leqslant \frac{1}{\delta^4} \sum \left(x - \frac{k}{n} \right)^4 p_{nk} \leqslant \frac{M}{n^2 \delta^4} = \frac{M_1}{n^2}, \qquad 0 \leqslant x \leqslant 1. \quad (13)$$

THEOREM 5 (Weierstrass). If $f \in C^*$, then for each $\epsilon > 0$ there exists a trigonometric polynomial T such that

$$| f(x) - T(x) | < \epsilon, \qquad x \in K. \quad (14)$$

Proof. We show that $\sigma_n(f) \to f$ for all $f \in C^* = C[K]$. This, too, follows from Theorem 3: We can take $f_1 = 1, f_2 = \cos x, f_3 = \sin x$, and

$$P_y(x) = 1 - \cos (y - x). \qquad \blacksquare$$

Functions of several variables can also be approximated by polynomials. We cannot discuss this subject at any length. However, we can easily prove

THEOREM 6. A continuous function $f(x_1, \cdots, x_s)$, defined on a compact subset A of the euclidean s-space, is approximable by polynomials in $x_1, \cdots, x_s$.

Proof. It is sufficient to prove this for the unit square $S: 0 \leqslant x_i \leqslant 1$, $i = 1, \cdots, s$, for the set A is contained in some square S', and by the theorem of Tietze (see [28]), we can extend f from A onto S', with the preservation of its continuity. Hence, it is sufficient to prove Theorem 6 for $S' = A$. But S' can be linearly mapped onto S.

For $A = S$, we have $B_n(f) \to f$, where the nth Bernstein polynomial B_n of f is defined by

$$B_n(f; x_1, \cdots, x_s) = \sum_{k_1=0}^{n} \cdots \sum_{k_s=0}^{n} f\left(\frac{k_1}{n}, \cdots, \frac{k_s}{n} \right) p_{nk_1}(x_1) \cdots p_{nk_s}(x_s). \quad (15)$$

In fact, the $B_n(f)$ are positive linear operators, and we have $B_n(f) \to f$ for each of the $2s + 1$ functions $1, x_i, x_i^2, i = 1, \cdots, s$.

For the function $P_y(x)$ of (1), we can take

$$P_y(x) = \sum_{i=1}^{s} (y_i - x_i)^2;$$

we have $P_y(x) \geqslant 0$ and $P_y(x) = 0$ if and only if $y = x$, with $y = (y_1, \cdots, y_s)$, $x = (x_1, \cdots, x_s)$. $\qquad \blacksquare$

4. The Theorem of Stone

The theorem of Stone [91] (which will not be used in this book) is more abstract in nature than our previous results. It deals with approximation by polynomials in an arbitrary set of functions. However, it can be derived from our "concrete" theorems. More specifically, it follows from Theorem 6 and some standard topological facts. We require, but only in this section, a knowledge of some basic facts about products of topological spaces. (See for example [28].)

Let A be a compact Hausdorff topological space, and $G = \{g\}$ be a family of continuous real functions on A. We consider arbitrary polynomials P with real coefficients in the functions g. In other words, the polynomials P are (finite) sums of the type

$$P(x) = \sum \alpha g_1(x)^{n_1} \cdots g_m(x)^{n_m} , \tag{1}$$

where $g_i \in G$, α are real and $n_i \geqslant 0$ are arbitrary integers. Under what conditions (to be formulated in terms of G) is it true that *each* continuous real function on A is approximable by the polynomials P? To answer this question, we shall use the following terminology: We shall say that G *distinguishes points of A* if for each pair of points $x_1 \neq x_2$ of A there is a function $g \in G$ with $g(x_1) \neq g(x_2)$. Obviously, this property is a *necessary* condition for our problem: If G does not distinguish points of A, there are two points $x_1 \neq x_2$ such that all g, and hence all P, take equal values at x_1, x_2. But there exist continuous functions f with $f(x_1) \neq f(x_2)$. Each such f is not approximable by the P.

THEOREM 7 (M. H. Stone).   If the set G distinguishes points of A, then each continuous real function is approximable by polynomials in the functions $g \in G$.

Proof.   Since each $g \in G$ is bounded, we can assume without loss of generality that for each g, we have $-1 \leqslant g(x) \leqslant 1$, $x \in A$.

For each $g \in G$, let $I_g = [-1, 1]$. Let $\pi = \prod_{g \in G} I_g$ be the topological product of the I_g; by Tychonoff's theorem, π is a compact Hausdorff space. A point y of π is a set $\{y\}_{g \in G}$, where $y_g \in I_g$. The projections of π onto the I_g are the mappings p_g given by $p_g(y) = y_g$.

We define a mapping of A into π by $y = \Phi(x) = \{g(x)\}_{g \in G}$. This function Φ is continuous because each function $p_g(\Phi(x)) = g(x)$ is continuous. It is one-to-one because G distinguishes points of A. Thus, Φ is a continuous one-to-one map from a compact space into a Hausdorff space; hence, Φ is a homeomorphism of A onto $B = \Phi(A)$.

Let $f \in C[A]$ and $\epsilon > 0$ be arbitrary. We wish to show that there exists a polynomial $P(x)$ of the form (1) for which $|f(x) - P(x)| < \epsilon$, $x \in A$. The last inequality is equivalent to $|f(\Phi^{-1}(y)) - P(\Phi^{-1}(y))| < \epsilon$, $y \in B$. Now the inverse map Φ^{-1} assigns to each point $y \in B$ the unique point $x \in A$ for which $g(x) = y_g$, $g \in G$. Thus, for each $g \in G$, $g(\Phi^{-1}(y)) = y_g$. We see that our

theorem will be proved if we can show that for each continuous real function
$F(y)$ on B, there is a polynomial $\bar{P}(y)$ of the form

$$\bar{P}(y) = \sum \alpha y_{g_1}^{n_1} \cdots y_{g_m}^{n_m}$$

for which

$$|F(y) - \bar{P}(y)| < \epsilon, \qquad y \in B. \tag{2}$$

We can further reduce our problem by means of Tietze's extension theorem.
The function $F(y)$ is continuous on the closed subset B of the compact Hausdorff
space π, and can be extended onto the whole of π. Hence, it is sufficient to
prove (2) for the case $B = \pi$, for each function $F \in C[\pi]$.

We consider two cases: (a) G *is finite*, say, $G = \{g_1, \cdots, g_m\}$. In this case,
π is the m-dimensional unit cube, and (2) follows from Theorem 6.

(b) G *is infinite.* Then we approximate $F(y)$ by a function that depends
only on finitely many y_g. We recall that a basis for the open sets of π is given
by open sets U of the following kind: A point z of π belongs to U if and only if a
finite number of its coordinates z_g are restricted to lie in some open sets in the I_g,
while the other coordinates remain unrestricted.

For each $y \in \pi$, there is an open basis element U_y, $y \in U_y$ such that
$|F(y') - F(y)| < \frac{1}{4}\epsilon$ for all $y' \in U_y$. Then, $|F(y'') - F(y')| < \frac{1}{2}\epsilon$ for
$y', y'' \in U_y$. The sets U_y, $y \in \pi$ cover π; hence, there is a finite subcover,
$U_{y_1}, \cdots, U_{y_p}$. Let Γ_i be the finite set of the g for which the coordinates z_g
of points $z \in U_{y_i}$ are restricted; let $\Gamma = \Gamma_1 \cup \cdots \cup \Gamma_p$. Then Γ is finite, say,
$\Gamma = \{g_1, \cdots, g_m\}$. For each $y \in \pi$, let $\hat{y}$ be the point of π with the coordinates
$\hat{y}_g = y_g$ if $g \in \Gamma$, $\hat{y}_g = 0$ if $g \notin \Gamma$. The point y must belong to some U_{y_i}; then,
also, $\hat{y} \in U_{y_i}$ by the definition of $\hat{y}$. It follows that

$$|F(y) - F(\hat{y})| < \frac{1}{2}\epsilon, \qquad y \in \pi. \tag{3}$$

The set of all points $\hat{y}$ is homeomorphic to the set of all points $\{y_{g_i}\}_{i=1}^m$ of the
product $\prod_1^m I_{g_i}$, and $F(\hat{y})$ can be considered to be a function on this set. By
case (a), for some polynomial $\bar{P}$ in the y_{g_i}, $|F(\hat{y}) - \bar{P}(y)| < \epsilon/2$. This, together
with (3), establishes the theorem.   ∎

Of course theorem 7 implies at once the special theorems of Weierstrass.
For instance, to obtain Theorem 5, we have only to observe that the functions
$\cos x$ and $\sin x$ separate points of K, and that polynomials in $\cos x$ and $\sin x$
are precisely the trigonometric polynomials. We stress again, that only real-
valued functions and polynomials with real coefficients are allowed in Theorem 7. For complex-valued functions, the theorem is *false*. Consider, for
example, the complex function z on the disc $D : |z| \leqslant 1$. It separates points of
D. However, uniform limits of polynomials $\lim P_n(z) = f(z)$ in D are not
arbitrary continuous functions. In fact, the polynomials P_n, and hence also the
limit functions f, are *analytic* for $|z| < 1$.

5. Notes

1.   The linear combinations of the functions Φ: 1, cos x, sin x, $\cdots$, cos nx, sin nx, $\cdots$, approximate each function from C^*, as we know from Theorem 5. If one of the functions in Φ is omitted, this property is lost (this is a common feature of all normed orthogonal sequences of functions). The situation with the sequence 1, x, x^2, $\cdots$, x^n, $\cdots$ is quite different. Here, we can omit almost all functions! More generally, we consider the functions 1, x^{p_1}, $\cdots$, x^{p_n}, $\cdots$ on [0, 1], where the p_n are real numbers subject only to the conditions $0 < p_1 < p_2 < \cdots$, $p_n \to \infty$. Müntz (see [1, p. 43], [10, p. 489]) proved that linear combinations of the functions approximate all continuous functions on [0, 1] if and only if

$$\sum p_n^{-1} = +\infty.$$

2.   The elements $f_1, \cdots, f_n, \cdots$ of a Banach space X form a *Schauder basis* if each element $f \in X$ has a unique expansion $f = \sum\limits_{n=1}^{\infty} a_n f_n$ (the convergence is in the norm of X). Schauder proved that the space $C[a, b]$ has a basis of this kind. Kreĭn, Mil'man, and Rutman [71] derived from this the existence of a sequence of algebraic polynomials $P_n(x)$ with the property that each continuous function $f(x)$ has a *unique uniformly convergent representation*

$$f(x) = \sum_{n=1}^{\infty} a_n P_n(x).$$

3.   There are several results about approximation by algebraic polynomials whose coefficients satisfy certain restrictions. We may assume, for example, that the coefficients are integers. In this case, the possibility of approximation depends upon the length of the interval [a, b]. If [a, b] does not contain integral points, approximation of a continuous function is unreservedly possible. If this interval is [0, 1], approximation of f is possible if and only if $f(0)$ and $f(1)$ are integers. As the length $b - a$ of the interval increases, the number of conditions that the values of f must satisfy increases indefinitely; finally, if $a - b \geqslant 4$, no nontrivial approximation by polynomials with integral coefficients is possible. The interested reader is referred to Hewitt and Zuckerman [57] and Gelfand [52]. Approximation by polynomials with *positive* coefficients has been discussed by Jurkat and Lorentz [59].

4.   Let X be an arbitrary Banach space, Φ a sequence of elements ϕ_i of X. Let us come back to the degree of approximation $E_n(f)$, 1(7). It is clear that these numbers form a decreasing sequence, and if all $f \in X$ are approximable by the ϕ_i, we have $E_n(f) \to 0$, $f \in X$. Bernstein has proved that this describes the sequence $E_n(f)$ completely. If X and Φ are given, and $\epsilon_n \geqslant 0$ is an arbitrary sequence that is monotone-decreasing to zero, then there is an element f of the space X with the property $E_n(f) = \epsilon_n$, $n = 0, 1, \cdots$ (see [4, p. 332], [10, p. 99]).

PROBLEMS

1. Each function $f \in C[0, \pi]$ can be uniformly approximated by even trigonometric polynomials.

2. A continuous function $f(x)$, $-\infty < x < +\infty$, which has a finite limit at infinity, $\lim\limits_{x \to \pm\infty} f(x)$, can be approximated on $(-\infty, +\infty)$ by rational functions of the form $(1 + x^2)^{-n} P_{2n}(x)$.

3. Let Q_n be a sequence of polynomials of degrees m_n, and $Q_n(x) \to f(x)$ uniformly on $[a, b]$, where f is not a polynomial. Then $m_n \to \infty$.

4. Under the same assumptions as in Problem 3, $mE_n \to 0$, where E_n is the set of points x with $Q_n(x) = f(x)$. (Brudnyĭ and Gopengaus.)

5. If $f \in C[0, 1]$ and $f(0) = f(1) = 0$, then the sequence of polynomials with integral coefficients

$$\sum_{k=1}^{n-1} \left[\binom{n}{k} f\left(\frac{k}{n}\right) \right] x^k (1 - x)^{n-k}$$

converges uniformly to $f(x)$.

6. A function $f \in C[0, 1]$ is approximable by polynomials with integral coefficients if and only if $f(0)$ and $f(1)$ are integers.

7. A function $f \in C[A]$ is approximable by polynomials 4(1) in $g \in G$ if the functions g distinguish all pairs of points of A that are distinguished by the function f.

8. If $A_1, \cdots, A_p$ are compact Hausdorff spaces and $A = A_1 \times \cdots \times A_p$, then each function $f(x_1, \cdots, x_p)$, continuous on A, is approximable by sums of the type $\sum_{k=1}^{n} f_{k1}(x_1) \cdots f_{kp}(x_p)$, where $f_{ki} \in C[A_i]$, $i = 1, \cdots, p$, $k = 1, \cdots, n$, (Dieudonné.)

9. Let $K_n(x, t)$ be continuous for $a \leqslant x$, $t \leqslant b$, let (a) $\int_a^b K_n(x, t)\, dt = 1$, (b) $\int_a^b | K_n(x, t) |\, dt \leqslant A$, $n = 1, 2, \cdots$, (c) $\int_{|x-t| \geqslant \delta} | K_n(x, t) |\, dt \to 0$ for $n \to \infty$, for each $\delta > 0$. Then $\int_a^b K_n(x, t)\, f(t)\, dt \to f(x)$ uniformly for each $f \in C[a, b]$.

10. Prove Theorem 1 by establishing that the polynomials

$$P_{2n}(f, x) = \lambda_n^{-1} \int_0^1 [1 - (t - x)^2]^n f(t)\, dt, \qquad \lambda_n = \int_{-1}^{+1} (1 - t^2)^n\, dt,$$

where $f \in C[0, 1]$, converge uniformly to $f(x)$ for $\delta \leqslant x \leqslant 1 - \delta$, for each $\delta > 0$. (Landau.)

11. If the function $f(x)$ is continuous on $[0, +\infty)$ and has limit zero for $x \to +\infty$, then $\lim\limits_{u\to\infty} S_u(x) = f(x)$ uniformly for $0 \leqslant x < +\infty$, if $S_u(x)$ is defined by

$$S_u(x) = e^{-ux} \sum_{k=0}^{\infty} \frac{(ux)^k}{k!} f\left(\frac{k}{u}\right), \qquad u > 0.$$

12. Prove Theorem 3 with condition 3(1) weakened, as indicated in the footnote on page 7.

[2]

Polynomials of Best Approximation

1. Existence of Polynomials of Best Approximation

Let X be a Banach space with real or complex scalars (see Chapter 1, Sec. 1), and let $x_1, \cdots, x_n$ be given vectors in X. We consider linear combinations ("polynomials") of the form $y = \sum_{i=1}^{n} a_i x_i$, where a_i are scalars. We recall that for each $x \in X$, the degree of approximation $E_n(x)$ of x by the polynomials y is

$$E(x) = E_n(x) = \inf_y \| x - y \| . \tag{1}$$

If the infimum is attained for some $y = y_0$, this y_0 is called a *linear combination of best approximation* or a *polynomial of best approximation* to x. In this chapter we shall discuss the questions of existence and uniqueness of polynomials of best approximation, and find their characteristic properties.

LEMMA 1. The set of all polynomials

$$y = \sum_{i=1}^{n} a_i x_i , \qquad \| y \| \leqslant M, \tag{2}$$

where the vectors $x_1, \cdots, x_n$ in X and the number $M \geqslant 0$ are given, is compact. In other words, from each sequence y_m, $m = 1, 2, \cdots$, of polynomials (2), one can extract a subsequence that converges to a polynomial of the same kind.

Proof. We can assume that the vectors $x_1, \cdots, x_n$ are linearly independent. Let a_{mi} be the coefficients of the polynomial y_m. In order to prove our statement, it is sufficient to show that the coefficients a_{mi} are bounded, that is, that there exists an M_1 such that $| a_{mi} | \leqslant M_1$ for all m and i. Indeed, for each i, in this case, the sequence a_{mi} would contain a convergent subsequence. By successive extraction of subsequences, we would obtain a subsequence y_{m_j} for which $a_{m_j i} \to a_i$ as $j \to \infty$ for each i. Then y_{m_j} would converge, in the norm of the space X, to $y = \sum_{1}^{n} a_i x_i$, with $\| y \| \leqslant M$.

Assume that M_1 does not exist. Replacing y_m, if necessary, by a subsequence (which we still denote by y_m), we can then assume that $\max_i | a_{mi} |$ is different from zero, tends to infinity as $m \to \infty$, and moreover that this maximum

16

is attained for each $m = 1, 2, \cdots$ for the same i; say, for $i = 1$. Put $z_m = y_m/a_{m1}$. Then $\| z_m \| \leqslant M/| a_{m1} | \to 0$ and z_m has the form

$$z_m = x_1 + b_{m2}x_2 + \cdots + b_{mn}x_n ,$$

with $| b_{mi} | \leqslant 1$. Since the b_{mi} are bounded, we can assume (replacing z_m, if necessary, by a subsequence) that $b_{mi} \to b_i$, $i = 2, \cdots, n$. Then

$$x_1 + b_2x_2 + \cdots + b_nx_n = 0,$$

which is impossible because the x_i are linearly independent. ∎

We can now prove

THEOREM 1.   For each set of vectors $x_1 , \cdots, x_n \in X$ and each $x \in X$, there is a polynomial of best approximation.

Proof.   There exists a sequence of polynomials y_m for which $\| x - y_m \| \to E(x)$. Since $\| y_m \| \leqslant \| x \| + \| x - y_m \|$ is bounded, we can apply Lemma 1 and obtain a sequence y_{m_j} with $\| y_{m_j} - y_0 \| \to 0$, where y_0 is some polynomial. Then $\| x - y_{m_j} \| \to \| x - y_0 \|$, and therefore $\| x - y_0 \| = E(x)$. ∎

For some Banach spaces X, we can claim that the polynomial of best approximation is *unique* for each system $x_1 , \cdots, x_n$. This is the case for *strictly convex spaces*, which are characterized by the following property:

$$x_1 \neq x_2 , \qquad \| x_1 \| = \| x_2 \| = 1, \qquad \alpha_1 , \alpha_2 > 0, \qquad \alpha_1 + \alpha_2 = 1$$

imply

$$\| \alpha_1 x_1 + \alpha_2 x_2 \| < 1. \tag{3}$$

If y_1 , y_2 are two polynomials for which

$$\| x - y_1 \| = \| x - y_2 \| = E(x) > 0,$$

then, using (3), we obtain a contradiction,

$$E(x) \leqslant \| x - \tfrac{1}{2}(y_1 + y_2) \| = \| \tfrac{1}{2}(x - y_1) + \tfrac{1}{2}(x - y_2) \| < E(x),$$

unless $y_1 = y_2$. It is easy to check that the spaces L^p, $1 < p < + \infty$, are strictly convex. (For this purpose, it is sufficient to know the cases of equality in the inequality of Minkowski; see [23, p. 146]). However, the important spaces L^1 and C do not have this property, and the uniqueness statement does not apply to them. For example, take the space $C[0, 1]$. Let $x(t) = 1$, and $x_1(t) = t$, $0 \leqslant t \leqslant 1$. We wish to approximate x by polynomials $y = a_1x_1$. For each y, $\| x - y \| \geqslant x(0) - y(0) = 1$. On the other hand, this lower bound for $\| x - y \|$ is attained by $y = a_1x_1$, $0 \leqslant a_1 \leqslant 2$. Hence, all these y are polynomials of best approximation.

2. Characterization of Polynomials of Best Approximation

The basic theorems about approximation in spaces C, in Sections 2 to 5, will be treated in the following setting. Let A be a compact Hausdorff topological space, $C[A]$ the space of all continuous complex or real functions f on A with the norm $\|f\| = \max_{x \in A} |f(x)|$. Let $\phi_1, \cdots, \phi_n$ be given (complex or real) continuous functions on A. A linear combination $P = \sum_{i=1}^{n} a_i \phi_i$ with complex or real coefficients a_i will be called a *polynomial* in the ϕ_i, as in Sec. 1. We shall treat the "complex case" and the "real case" together; although some formulas in Secs. 2 to 5 are simpler for the "real case," there is no essential simplification of the proofs. Only in Sec. 6 will we encounter theorems that are restricted to the "real case."

For a complex number $\alpha = a + bi$ with real part $a = \operatorname{Re} \alpha$ and imaginary part $b = \operatorname{Im} \alpha$, we write $\bar{\alpha} = a - bi$. Then

$$(\overline{\alpha\beta}) = \bar{\alpha}\bar{\beta}, \qquad \alpha\bar{\alpha} = |\alpha|^2, \qquad \operatorname{Re} \alpha = \tfrac{1}{2}(\alpha + \bar{\alpha}).$$

We note the following identity, which will be useful:

$$|\alpha + \beta|^2 = (\alpha + \beta)(\bar{\alpha} + \bar{\beta}) = |\alpha|^2 + 2\operatorname{Re} \alpha\bar{\beta} + |\beta|^2. \tag{1}$$

The desired characterization of polynomials of best approximation is given by the following theorem.

THEOREM 2 (Kolmogorov [61]). P is a polynomial of best approximation for a continuous function f if and only if for each polynomial Q,

$$\max_{x \in A_0} \operatorname{Re} \{[f(x) - P(x)] \overline{Q(x)}\} \geqslant 0, \tag{2}$$

where A_0 denotes the set (which depends on f and P) of all points $x \in A$ for which $|f(x) - P(x)| = \|f - P\|$.

For the "real case," (2) takes the form

$$\max_{x \in A_0} \{[f(x) - P(x)] Q(x)\} \geqslant 0. \tag{3}$$

The last condition means that the relation $[f(x) - P(x)] Q(x) < 0$ cannot be satisfied for all $x \in A_0$; in other words, that $f(x) - P(x)$ and $Q(x)$ cannot be of opposite signs for all $x \in A_0$. Relation (3) means roughly that A_0 is large and that $f(x) - P(x)$ changes sign frequently on A_0. Later, in Theorems 5 and 9, we shall see how this vague statement can be made more precise.

Proof. Assume first that $P(x)$ is a polynomial of best approximation. Let $\|f - P\| = E$. If (2) is not true, there exists a polynomial $Q(x)$ such that

$$\max_{x \in A_0} \operatorname{Re} \{[f(x) - P(x)] \overline{Q(x)}\} = -2\epsilon$$

for some $\epsilon > 0$. By the continuity of the function, there exists an open subset G of A, $G \supset A_0$, such that

$$\text{Re}\{[f(x) - P(x)]\overline{Q(x)}\} < -\epsilon, \qquad x \in G.$$

Let us see how f is approximated by the polynomial $P_1 = P - \lambda Q$, where $\lambda > 0$ is small. Let M denote the maximum of $|Q(x)|$ on A. First we assume that $x \in G$. By (1),

$$|f(x) - P_1(x)|^2 = |[f(x) - P(x)] + \lambda Q(x)|^2$$
$$= |f(x) - P(x)|^2 + 2\lambda \text{Re}\{[f(x) - P(x)]\overline{Q(x)}\} + \lambda^2 |Q(x)|^2$$
$$< E^2 - 2\lambda\epsilon + \lambda^2 M^2.$$

If we take $\lambda < M^{-2}\epsilon$, then $\lambda^2 M^2 < \lambda\epsilon$, and we obtain

$$|f(x) - P_1(x)|^2 < E^2 - \lambda\epsilon, \qquad x \in G. \tag{4}$$

To estimate $f(x) - P_1(x)$ for $x \notin G$, we note that the complement of G is a closed set $F \subset A$, and that on F, $|f(x) - P(x)| < E$. Hence, for some $\delta > 0$, $|f(x) - P(x)| < E - \delta$, $x \in F$. If we take λ so small that $\lambda < (2M)^{-1}\delta$, we shall have

$$|f(x) - P_1(x)| \leqslant |f(x) - P(x)| + \lambda |Q(x)| \leqslant E - \delta + \tfrac{1}{2}\delta = E - \tfrac{1}{2}\delta, \qquad x \in F. \tag{5}$$

From (4) and (5) we see that for all sufficiently small positive values of λ, P_1 approximates f better than P. Hence, the condition (2) is necessary.

To show that (2) is also sufficient, assume that (2) holds for each Q. Taking an arbitrary polynomial P_1, we see that there is a point $x_0 \in A_0$ such that for $Q = P - P_1$,

$$\text{Re}\{[f(x_0) - P(x_0)]\overline{Q(x_0)}\} \geqslant 0.$$

Then, by (1),

$$|f(x_0) - P_1(x_0)|^2 = |f(x_0) - P(x_0)|^2 + 2\text{Re}\{[f(x_0) - P(x_0)]\overline{Q(x_0)}\} + |Q(x_0)|^2$$
$$\geqslant |f(x_0) - P(x_0)|^2 = \|f - P\|^2,$$

since $x \in A_0$. Hence, P_1 cannot approximate f with an error less than $\|f - P\|$, and P must be a polynomial of best approximation. ∎

3. Applications of Convexity

Convexity considerations are useful in the theory of approximation. In this section we shall be able to use for this purpose the properties of convex sets in finite dimensional spaces.

A subset B of a real n-dimensional euclidean space R_n is *convex* if with every two points $x, y \in B$, it contains the segment xy, that is, all points $\alpha x + \beta y$ for which $\alpha, \beta \geqslant 0$ and $\alpha + \beta = 1$. We shall list some well-known properties of convex sets (consult [22] for proofs). If B is convex, $x^{(i)} \in B$, $p_i \geqslant 0$, $i = 1, \cdots, r$, and $\sum p_i = 1$, then $\sum p_i x^{(i)} \in B$. The intersection of any collection of convex sets is convex. The closure of a convex set is convex. For each set $B \subset R_n$, there exists a *smallest* convex set B_c that contains B; this set B_c is called the *convex hull* of B.

A *hyperplane* in R_n is the set of points $x = (x_1, \cdots, x_n)$ satisfying $\sum_1^n a_k x_k = a_0$, where a_k are real and not all a_k, $k = 1, \cdots, n$, are zero. Under the same assumptions, the set H of points for which $\sum_1^n a_k x_k \leqslant a_0$ is called a *closed half-space*. A compact convex set B is the intersection of all closed half-spaces H that contain B:

$$B = \bigcap_{H \supset B} H. \tag{1}$$

The following simple lemma will be useful.

LEMMA 2.   The convex hull of a set $B \subset R_n$ consists of all points x of the form

$$x = \sum_{i=1}^r p_i x^{(i)}, \qquad x^{(i)} \in B, \, p_i > 0, \, i = 1, \cdots, r; \quad \sum_1^r p_i = 1, \tag{2}$$

where $r \leqslant n + 1$.

Proof.   Let B' denote the set of all points x that are representable in the form (2) for some r, where $r > n + 1$ is permitted. Clearly, $B' \subset B_c$. On the other hand, B' is convex: If x and y are of the form (2) (with perhaps different r), then also $\alpha x + \beta y$, $\alpha > 0$, $\beta > 0$, $\alpha + \beta = 1$ is of this form. Since $B' \supset B$, we see that $B' = B_c$.

Now let $x \in B_c$ be given, and let r denote the smallest integer for which a representation (2) is possible. We have to prove that $r \leqslant n + 1$.

Assume that $r > n + 1$. Then the $r - 1 > n$ vectors $x^{(2)} - x^{(1)}, \cdots, x^{(r)} - x^{(1)}$ are linearly dependent, and there is a relation

$$\sum_{i=1}^{r-1} \beta_i [x^{(i+1)} - x^{(1)}] = \sum_{i=1}^r \alpha_i x^{(i)} = 0, \qquad \sum \alpha_i = 0,$$

where not all β_i and not all α_i are zero. Thus, we have in addition to (2),

$$x = \sum_{i=1}^r (p_i + \lambda \alpha_i) x^{(i)}.$$

We want to choose λ so that $q_i = p_i + \lambda \alpha_i \geqslant 0$ for all i and so that $q_i = 0$ for at least one i. There are α_i that satisfy $\alpha_i < 0$. It is easy to see that $\lambda = \min \{p_i / |\alpha_i|; \alpha_i < 0\} > 0$ has the required properties. Since $\sum q_i = 1$, we have obtained a representation of type (2) with less than r terms. This is a contradiction. ∎

We note some useful corollaries. Let B be compact. Then we can imitate the proof of Lemma 1. Using Lemma 2 and the compactness of B, we prove that each sequence of points of B_c has a subsequence that converges to a point of B_c. Therefore, *the convex hull of a compact set is compact*. Using this and (1), we obtain, if B is compact,

$$B_c = \bigcap_{H \supset B_c} H = \bigcap_{H \supset B} H. \tag{3}$$

All the foregoing considerations apply to convexity questions in the n-dimensional *complex* space R_n, of points $z = (z_1, \cdots, z_n)$ with complex coordinates $z_k = x_k + iy_k$, which is isomorphic to the $2n$-dimensional, real euclidean space. In this case, the restriction upon r in Lemma 2 becomes $r \leqslant 2n + 1$. The defining inequality of a half-space now is

$$\sum_{k=1}^{n} (a_k x_k + b_k y_k) \leqslant a_0.$$

We can combine the complex and the real cases by writing this in the form

$$\mathrm{Re} \left\{ \sum_{k=1}^{n} \bar{c}_k z_k \right\} \leqslant a_0, \tag{4}$$

where c_k are arbitrary complex (real) numbers, not all zero, and a_0 is real.

In the discussion that follows, $C[A]$ will be the space of complex or real continuous functions on a compact Hausdorff topological space A, $\Phi = \{\phi, \cdots, \phi_n\}$, a set of n elements of $C[A]$. For given $f \in C[A]$ and $P = \sum a_k \phi_k$, we should like to decide whether P is a polynomial of best approximation for f. As before, we denote by A_0 the set of all $x \in A$ with the property

$$|f(x) - P(x)| = \|f - P\|.$$

We begin by giving a new form to the condition 2(2) of Theorem 2:

The origin in R_n belongs to the convex hull of the set B of points ⎱
$$z = ((f(x) - P(x))\overline{\phi_1(x)}, \cdots, (f(x) - P(x))\overline{\phi_n(x)}), \qquad x \in A_0. \tag{5}$$
⎰

The set A_0 is compact, and $x \to z$ is a continuous map of A into R_n. Hence, B (the image of A_0 in R_n) is also compact. Now, by (3), the point 0 belongs to B_c if and only if it belongs to each closed half-space H which contains all points z of B. If H is given by (4), then $0 \in H$ means that $a_0 \geqslant 0$. Therefore, condition (5) is equivalent to the statement that if for some c_i and a_0,

$$\mathrm{Re} \left\{ \sum_{i=1}^{n} \bar{c}_i (f(x) - P(x)) \overline{\phi_i(x)} \right\} \leqslant a_0 \qquad \text{for all} \qquad x \in A_0,$$

then $a_0 \geqslant 0$. Putting $Q = \sum c_i \phi_i$, we see that this is equivalent to 2(2).

Combining condition (5) with Lemma 2, we obtain

THEOREM 3. A polynomial P is a polynomial of best approximation for $f \in C[A]$ if and only if there are r points $x_1, \cdots, x_r \in A_0$, and r numbers $p_1 > 0, \cdots, p_r > 0, \sum p_k = 1$ $(r \leqslant 2n + 1$ in the complex case and $r \leqslant n + 1$ in the real case), for which

$$\sum_{k=1}^{r} p_k[f(x_k) - P(x_k)] \overline{\phi_i(x_k)} = 0, \qquad i = 1, \cdots, n, \tag{6}$$

or equivalently,

$$\sum_{k=1}^{r} p_k[f(x_k) - P(x_k)] \overline{Q(x_k)} = 0 \qquad \text{for each polynomial } Q. \quad \blacksquare \tag{7}$$

As a corollary we have

THEOREM 4. If P is a polynomial of best approximation for f on A, then P is also a polynomial of best approximation for f on a certain finite subset of A, which consists of r ($r \leqslant 2n + 1$ or $r \leqslant n + 1$) points.

We shall say that a function μ on A has the *finite support* $S = \{x_1, \cdots, x_r\}$, where the x_k, $k = 1, \cdots, r$, are distinct points of A, if $\mu(x) = 0$ on $A \setminus S$ and if $\mu(x_k) \neq 0$, $k = 1, \cdots, r$. A *signature* σ is a function of finite support S whose values $\sigma(x_k)$ at the points $x_k \in S$ are (complex or real) signs, that is, numbers of absolute value 1. A signature σ with support $S = \{x_1, \cdots, x_r)$ is an *extremal signature* (with respect to the system Φ) if there exists a function μ with support S for which sign $\mu(x_k) = \sigma(x_k)$, $k = 1, \cdots, r$, and

$$\sum_{k=1}^{r} \mu(x_k) \overline{Q(x_k)} = 0 \qquad \text{for all polynomials } Q. \tag{8}$$

The following theorem shows that the notion of extreme signature is useful.

THEOREM 5 (Rivlin and Shapiro [86]). A polynomial P, not identically equal to the function $f \in C[A]$, is a polynomial of best approximation for f if and only if there exists an extremal signature σ with support $S = \{x_1, \cdots, x_r\} \subset A_0$ for which $r \leqslant 2n + 1$ (or $r \leqslant n + 1$) and

$$\text{sign}\,[f(x_k) - P(x_k)] = \sigma(x_k), \qquad k = 1, \cdots, r. \tag{9}$$

Proof.

Necessity. For a polynomial of best approximation P, we have (7). We define $\sigma(x_k) = \text{sign}\,[f(x_k) - P(x_k)]$, $\mu(x_k) = p_k[f(x_k) - P(x_k)]$, $k = 1, \cdots, r$, and put $\mu(x) = \sigma(x) = 0$ for $x \neq x_k$, $k = 1, \cdots, r$. Then (7) shows that σ is an extremal signature, and (9) holds.

Sufficiency. Let the conditions of the theorem be satisfied and let μ be the function with support S that satisfies (8). Then, for each polynomial P_1,

$$\sum [\overline{P(x_k) - P_1(x_k)}]\, \mu(x_k) = 0,$$

and therefore

$$\|f - P_1\| \sum_{k=1}^{r} |\mu(x_k)| \geqslant \left| \sum [\overline{f(x_k) - P_1(x_k)}]\, \mu(x_k) \right|$$

$$= \left| \sum [\overline{f(x_k) - P(x_k)}]\, \mu(x_k) \right|$$

$$= \sum |f(x_k) - P(x_k)|\, |\mu(x_k)|$$

$$= \|f - P\| \sum_{k=1}^{r} |\mu(x_k)|. \tag{10}$$

Hence,

$$\|f - P_1\| \geqslant \|f - P\|. \qquad \blacksquare$$

The computation (10) is useful in a more general case. Assume that σ *is an extremal signature with support* $S = \{x_1, \cdots, x_r\}$, *and that for some polynomial* P *and some* $\delta \geqslant 0$,

$$\text{sign}\,[f(x_k) - P(x_k)] = \sigma(x_k), \quad |f(x_k) - P(x_k)| > \delta, \quad k = 1, \cdots, r. \tag{11}$$

Then

$$E_n(f) \geqslant \delta. \tag{12}$$

Indeed, we have now $\|f - P_1\| \geqslant \delta$ for each polynomial P_1.

4. Chebyshev Systems

An important property of the functions $1, x, x^2, \cdots, x^n$ is that a polynomial $a_0 + a_1 x + \cdots + a_n x^n$ that does not vanish identically can have no more than n zeros on an interval $[a, b]$. The functions $1, z, z^2, \cdots, z^n$ enjoy the same property on each subset of the complex plane. Also, the trigonometric functions $1, \cos x$, $\sin x, \cdots, \cos nx, \sin nx$ have this property. A trigonometric polynomial with real or complex coefficients

$$T(x) = a_0 + a_1 \cos x + b_1 \sin x + \cdots + a_n \cos nx + b_n \sin nx,$$

which is not identically zero, cannot have more than $2n$ distinct zeros on the circle K. For the proof, consider the one-to-one correspondence between the points of the circle $|z| = 1$ and values of $x \bmod 2\pi$ given by $e^{ix} = z$. Using Euler's formulas,

$$\cos x = \frac{1}{2}(e^{ix} + e^{-ix}), \quad \sin x = \frac{1}{2i}(e^{ix} - e^{-ix}),$$

we obtain

$$T(x)\, e^{inx} = P(z), \tag{1}$$

where P is a polynomial of degree $2n$. If T has more than $2n$ zeros, different mod 2π, then P has more than $2n$ distinct zeros on $|z| = 1$; hence, P is identically zero, and from (1) we see that T is also identically zero.

We shall adopt the following definition: A set of continuous, complex, or real functions ϕ_1, $\cdots$, ϕ_n on a compact Hausdorff topological space A is a *Chebyshev system* if the following conditions are satisfied:

(a) A contains at least n points.

(b)   Each polynomial $P = a_1\phi_1 + \cdots + a_n\phi_n$, which does not have all coefficients a_i equal to zero, has at most $n - 1$ distinct zeros on A.

It follows in particular from this definition that the functions of a Chebyshev system are linearly independent. The three systems mentioned at the beginning of this section furnish important examples of Chebyshev systems. We leave it to the reader to prove that 1, $\cos x$, $\cdots$, $\cos nx$, and $\sin x$, $\cdots$, $\sin nx$ are two Chebyshev systems on $[0, \pi]$.

Condition (b) of the definition can be expressed also in the following form:

1.   If x_1, $\cdots$, x_n are distinct points of A, then the system of n equations with n unknowns a_1, $\cdots$, a_n,

$$a_1\phi_1(x_k) + a_2\phi_2(x_k) + \cdots + a_n\phi_n(x_k) = 0, \qquad k = 1, \cdots, n, \tag{2}$$

has only the obvious solution $a_1 = \cdots = a_n = 0$.

Using well-known facts about systems of linear equations, we see that this is also equivalent to each of properties 2 and 3.

2.   If x_1, $\cdots$, x_n are n distinct points of A, the determinant

$$D(x_1, \cdots, x_n) = \begin{vmatrix} \phi_1(x_1) & \cdots & \phi_n(x_1) \\ \cdots & \cdots & \cdots \\ \phi_1(x_n) & \cdots & \phi_n(x_n) \end{vmatrix} \tag{3}$$

is not zero.

3.   If x_1, $\cdots$, x_n are distinct points of A and c_1, $\cdots$, c_n are arbitrary numbers, then the system of equations

$$a_1\phi_1(x_k) + \cdots + a_n\phi_n(x_k) = c_k, \qquad k = 1, \cdots, n, \tag{4}$$

has a unique solution for the a_1, $\cdots$, a_n.

If we put $P = a_1\phi_1 + \cdots + a_n\phi_n$, conditions (4) read $P(x_k) = c_k$, $k = 1, \cdots, n$. We can call this P an *interpolating polynomial* with prescribed values c_k at the points x_k. Thus, statement 3 means that an interpolating polynomial P exists and is unique. If the number of the given points x_k and values c_k is less than n, then an interpolating polynomial also exists, but is not unique.

In the remainder of this section we shall consider *real* Chebyshev systems. If a compact space A is given, does there exist a real Chebyshev system for A? The answer is, in general, "no." We can prove that a space A has no real Chebyshev system with $n \geqslant 2$ if A contains three nonintersecting arcs emanating from a common point a. For example, there is no Chebyshev system on a two-dimensional square.

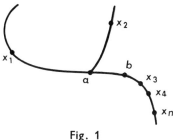

Fig. 1

In fact, assume that a Chebyshev system exists for some $n \geqslant 2$; we select the points $x_1, \cdots, x_n$ as indicated (Fig. 1). Put

$$g(x, y) = D(x, y, x_3, \cdots, x_n);$$

by statement 2, $g(x, y) \neq 0$ as long as $x, y, x_3, \cdots, x_n$ remain distinct. We move the points x, y continuously in A in the following manner: First, x moves from x_1 to b; then y moves from x_2 through a to x_1; finally, x moves from b through a to x_2. The function g will change continuously and remain different from zero; hence, $g(x_1, x_2)$ and $g(x_2, x_1)$ will have the same sign. But this is impossible, since $g(x_2, x_1) = -g(x_1, x_2)$.

The following statement holds for a real Chebyshev system $\phi_1, \cdots, \phi_n$, if $A = [a, b]$ or $A = K$.

4. If $x_1, \cdots, x_{n-1}$ are $n-1$ distinct points of A, then there exists a polynomial $D(x)$ that vanishes exactly at the points x_k and changes sign at each of these points (except when the point x_k coincides with a or b).

If the ϕ_i are the powers $1, x, \cdots, x^{n-1}$ on $[a, b]$, we can simply take

$$D(x) = (x - x_1) \cdots (x - x_{n-1}).$$

In the general case, we must modify this formula.

We take (see (3))

$$D(x) = D(x, x_1, x_2, \cdots, x_{n-1}).$$

Clearly, $D(x)$ vanishes for $x = x_1, \cdots, x_{n-1}$. If x is different from each of these points, $D(x) \neq 0$ according to statement 2. Hence, $D(x)$ keeps a constant sign on each interval between any two points x_k. Let us show, for example, that D changes sign at x_1 (if x_1 is different from a, b). Take $h > 0$ so small that the interval $[x_1 - h, x_1 + h]$ is contained in A and does not contain the points $x_2, \cdots, x_{n-1}$ (and the points a, b if $A = [a, b]$). Consider the function of t:

$$g(t) = D(x_1 - h + t, x_1 + t, x_2, \cdots, x_{n-1}),$$

which is continuous and different from zero for $0 \leqslant t \leqslant h$. Clearly, $g(t)$ has a constant sign in this interval. But

$$g(0) = D(x_1 - h), \qquad g(h) = D(x_1, x_1 + h, x_2, \cdots, x_{n-1}) = - D(x_1 + h).$$

This shows that $D(x_1 - h)$ and $D(x_1 + h)$ are of opposite sign and that D changes sign at x_1.

Property 4 is often used in the proof of Chebyshev's theorem (Theorem 9, Sec. 6). As a corollary of statement 4, we note: Each real Chebyshev system on K consists of an odd number of functions. Indeed, the continuous function $D(x)$ can have only an even number of changes of sign on K.

5. Uniqueness of Polynomials of Best Approximation

Here we shall prove that, for a Chebyshev system, each continuous function has *only one* polynomial of best approximation.

We shall need the following lemma.

LEMMA 3.   Let $\Phi = \{\phi_1, \cdots, \phi_n\}$ be a Chebyshev system of complex or real functions on a compact Hausdorff space A that contains at least $n + 1$ points, and let P be a polynomial of best approximation for a continuous function f. Then the set A_0 of all points $x \in A$, for which

$$|f(x) - P(x)| = \|f - P\| = E,$$

contains at least $n + 1$ points.

Proof.   Suppose that $A_0 = \{x_1, \cdots, x_s\}$, where $s \leqslant n$. Since there are points x with $|f(x) - P(x)| < E$, we must have $E > 0$. By Sec. 4, statement 3, there is a polynomial Q such that

$$Q(x_k) = - [f(x_k) - P(x_k)], \qquad k = 1, \cdots, s.$$

Then

$$\max_{x \in A_0} \mathrm{Re}\, \{[f(x) - P(x)] \overline{Q(x)}\} = \max_{k \leqslant s} \{- |f(x_k) - P(x_k)|^2\}$$

$$= - E^2 < 0,$$

which contradicts Theorem 2.   ∎

THEOREM 6.   For a Chebyshev system, there is a unique polynomial of best approximation for each continuous function.

Proof. If A has exactly n points, then by Sect. 4, statement 3, each function on A is equal to a polynomial P, and this polynomial is unique. Hence, we can assume that A consists of at least $n + 1$ points; then Lemma 3 is applicable.

Assume that for a continuous function $f(x)$ there are two polynomials of best approximation, $P(x)$ and $P_1(x)$:

$$\| f - P \| = \| f - P_1 \| = E.$$

Let

$$Q(x) = \tfrac{1}{2}(P(x) + P_1(x));$$

then

$$\| f - Q \| = \| \tfrac{1}{2}(f - P) + \tfrac{1}{2}(f - P_1) \| \leqslant \tfrac{1}{2}\| f - P \| + \tfrac{1}{2}\| f - P_1 \| = E.$$

On the other hand, $\| f - Q \| \geqslant E$, so that Q is also a polynomial of best approximation. By Lemma 3, there are at least $n + 1$ points x for which

$$| f(x) - Q(x) | = E.$$

At each such point x, for the complex numbers

$$\alpha = f(x) - P(x), \qquad \alpha_1 = f(x) - P_1(x),$$

we have $| \alpha + \alpha_1 | = 2E$, $| \alpha | \leqslant E$, $| \alpha_1 | \leqslant E$. But this is possible only if $\alpha = \alpha_1$. Thus, $P(x) = P_1(x)$ for at least $n + 1$ points of A. Since Φ is a Chebyshev system, the polynomials $P(x)$ and $P_1(x)$ are identical.   ∎

It is very remarkable that the converse of Theorem 6 is true.

THEOREM 7 (A. Haar [55]).   Let $\phi_1, \cdots, \phi_n$ be continuous functions defined on a compact set A that contains at least n points. If each continuous function has only one polynomial of best approximation, then $\phi_1, \cdots, \phi_n$ is a Chebyshev system.

Proof.   Clearly, the functions $\phi_1, \cdots, \phi_n$ are linearly independent. We shall assume that our system is not a Chebyshev system, and construct a function f_1, which has several polynomials of best approximation.

By assumption, there exist n distinct points $x_1, \cdots, x_n$ of A such that the system of linear equations

$$a_1\phi_1(x_k) + \cdots + a_n\phi_n(x_k) = 0, \qquad k = 1, \cdots, n \tag{1}$$

has nonzero solutions $a_1, \cdots, a_n$. We select a solution of this kind and put

$$P_1(x) = a_1\phi_1(x) + \cdots + a_n\phi_n(x).$$

The transposed system

$$c_1\phi_i(x_1) + \cdots + c_n\phi_i(x_n) = 0, \qquad i = 1, \cdots, n \tag{2}$$

also has nonzero solutions $c_1, \cdots, c_n$. Let $c_1, \cdots, c_n$ be such a solution. Some of the c_k may be zero, but the set N of k for which $c_k \neq 0$ is not empty.

According to Tietze's extension theorem, there exists a continuous function f_0 with the properties:

(a) $f_0(x_k) = \overline{c_k}/|c_k|,\ k \in N$.

(b) $|f_0(x)| \leqslant 1$ for $x \in A$.

LEMMA 4. For each continuous function f_0 satisfying conditions (a) and (b), we have $E_n(f_0) = 1$.

By (b), the polynomial $P = 0$ approximates f_0 with an error not exceeding 1; hence, $E_n(f_0) \leqslant 1$. We have to show that $E_n(f_0) < 1$ leads to a contradiction. Assume that there is a polynomial $P = \sum\limits_{i=1}^{n} b_i \phi_i$ such that $|f_0(x) - P(x)| < 1$ for all $x \in A$. In particular, by 2(1),

$$|f_0(x_k) - P(x_k)|^2 = |f_0(x_k)|^2 - 2\,\mathrm{Re}\,\{P(x_k)\overline{f_0(x_k)}\} + |P(x_k)|^2 < 1.$$

For $k \in N$, by (a), $|f_0(x_k)| = 1$; hence,

$$\mathrm{Re}\,\{P(x_k)\overline{f_0(x_k)}\} > 0, \qquad k \in N. \tag{3}$$

In other words, $\mathrm{Re}\,\{c_k P(x_k)\} > 0$ for $k \in N$. For all other values of k, this is equal to zero, since $c_k = 0$. Thus,

$$\mathrm{Re}\left\{\sum_{k=1}^{n} c_k P(x_k)\right\} > 0. \tag{4}$$

This, however, contradicts the equation

$$\sum_{k=1}^{n} c_k P(x_k) = \sum_{k=1}^{n} c_k \sum_{i=1}^{n} b_i \phi_i(x_k) = \sum_{i=1}^{n} b_i \sum_{k=1}^{n} c_k \phi_i(x_k) = 0,$$

which follows from (2).   ∎

To complete the proof of the theorem, we put

$$f_1(x) = (1 - M^{-1}|P_1(x)|)f_0(x),$$

where M is the maximum of $|P_1(x)|$ on A (we have $M > 0$, since the functions ϕ_i are linearly independent). By (1), $P_1(x_k) = 0$, $k = 1, \cdots, n$, and it is easy to see that $f_1(x)$ satisfies both conditions (a) and (b). Therefore, by Lemma 4, $E_n(f_1) = 1$. Finally, for all sufficiently small $\lambda > 0$, $\lambda P_1(x)$ is a polynomial of best approximation for the function f_1. In fact,

$$|f_1(x) - \lambda P_1(x)| \leqslant |f_1(x)| + \lambda|P_1(x)| \leqslant 1 - M^{-1}|P_1(x)| + \lambda|P_1(x)| \leqslant 1,$$

if $\lambda \leqslant M^{-1}$.   ∎

6. Chebyshev's Theorem

For Chebyshev systems $\Phi = \{\phi_1, \cdots, \phi_n\}$ on an interval or the circle, one can give simpler characterizations of polynomials of best approximation than those provided by Theorems 2 and 5. We consider the real case. Our first purpose will be to characterize extremal signatures (see Sec. 3). We have:

1. Each extremal signature of a Chebyshev system Φ on a compact Hausdorff topological space A has a support $S = (x_1, \cdots, x_r)$ of at least $n + 1$ points.

For if $r \leqslant n$, then (see Sec. 4) there is a polynomial Q for which $Q(x_1) = 1$, $Q(x_i) = 0$, $i > 1$. Then 3(8) imply $\mu(x_1) = 0$, so that an extremal signature cannot exist.

2. For each set of $n + 1$ distinct points $x_1, \cdots, x_{n+1} \in A$, there is a unique (up to multiplication with ± 1) extremal signature.

We consider the equation

$$\sum_{i=1}^{n+1} a_i Q(x_i) = 0, \qquad \text{for all polynomials } Q, \tag{1}$$

or (which is the same) the system

$$\sum_{i=1}^{n+1} a_i \phi_k(x_i) = 0, \qquad k = 1, \cdots, n. \tag{2}$$

Since the rank of the matrix $\| \phi_k(x_i) \|$ is n, the set of solutions $(a_1, \cdots, a_{n+1})$ is a one-dimensional space. Moreover, for a nonzero solution, no a_i can be equal to zero. For a solution of this kind, we put $\mu(x_i) = a_i$, $i = 1, \cdots, n + 1$, $\mu(x) = 0$ elsewhere. Then μ satisfies 3(8), and the function $\sigma(x) = \text{sign } \mu(x)$ is an extremal signature, with values $\sigma(x_i) = \pm 1$, $i = 1, \cdots, n + 1$. Any other function μ' that satisfies 3(8) is a multiple of μ, and the corresponding σ' satisfies $\sigma' = \pm \sigma$. ∎

Let $S = \{x_1, \cdots, x_{n+1}\} \subset A$, $A = [a, b]$ or $A = K$. Two points x_i, x_j of S are *adjacent*, if there is an interval in A, which contains x_i, x_j and no other points of S. A signature σ with support S will be called a *Chebyshev alternation* if the signs of σ are opposite in adjacent points. For example, if $A = [a, b]$ and

$$a \leqslant x_1 < \cdots < x_{n+1} \leqslant b,$$

then a function σ with support S is a Chebyshev alternation if and only if $\sigma(x_i) = (-1)^i$ or $\sigma(x_i) = (-1)^{i+1}$, $i = 1, \cdots, n + 1$.

THEOREM 8.   For a real Chebyshev system Φ on $A = [a, b]$ or $A = K$, extremal signatures with support of $n + 1$ points are identical with Chebyshev alternations.

Proof. We have to show that the signs of each nonzero solution $(a_1, \cdots, a_{n+1})$ of the system (2) are a Chebyshev alternation. First let $n = 1$; then ϕ_1 is of a constant sign on A and (2) consists of one equation, $a_1\phi_1(x_1) + a_2\phi_1(x_2) = 0$. In this case, sign $a_1 = -$ sign a_2.

Next let $n \geqslant 2$. There are at least three points x_i. We assume that at two adjacent points x_i, the a_i have the same sign; say, the positive sign. Relabeling the points x_i if necessary, we arrive at the following situation: The points x_{q-1} and x_q, and also x_q and x_{q+1}, are adjacent to each other, and $a_{q-1} > 0$, $a_q > 0$. We can select $\alpha_{q-1} > 0$, $\alpha_{q+1} > 0$ in such a way that

$$a_{q-1}\alpha_{q-1} + a_{q+1}\alpha_{q+1} > 0.$$

By 4(3), there exists a polynomial Q with values $Q(x_{q-1}) = \alpha_{q-1}, Q(x_{q+1}) = \alpha_{q+1}$, $Q(x_i) = 0$ for $i \neq q - 1, q, q + 1$. From (1),

$$a_{q-1}\alpha_{q-1} + a_q Q(x_q) + a_{q+1}\alpha_{q+1} = 0.$$

Therefore, $a_q Q(x_q) < 0$ and $Q(x_q) < 0$. Thus, the polynomial Q has a zero between x_{q-1} and x_q and also between x_q and x_{q+1}, altogether giving at least n zeros. This is a contradiction. ∎

Combining Theorems 5 and 8, we obtain at once

THEOREM 9 (Chebyshev). For a Chebyshev system on $A = [a, b]$ or $A = K$, let P be a polynomial that is not identical with the function $f \in C[A]$. Then P is a polynomial of best approximation for f if and only if there are $n + 1$ distinct points of A where $|f(x) - P(x)|$ attains its maximum and the signs of $f(x) - P(x)$ form a Chebyshev alternation. ∎

From the remark made in Sec. 3 after the proof of Theorem 5, we derive

THEOREM 10 (de la Vallée-Poussin). Under the assumptions of Theorem 9, let $|f(x_i) - P(x_i)| \geqslant \delta$, $i = 1, \cdots, n + 1$, and let the signs of $f(x_i) - P(x_i)$ form a Chebyshev alternation. Then

$$E_n(f) \geqslant \delta. \qquad (3)$$

7. Chebyshev Polynomials

Cases when there is an explicit formula for the degree of approximation $E_n(f)$, or for the polynomial of best approximation for f, are exceptional and are of special interest. Here and in Sec. 8, we shall consider two examples.

We begin with the approximation of x^n on $[-1, +1]$ by polynomials of degree $n - 1$. We wish to find a polynomial $a_{n-1}x^{n-1} + \cdots + a_1 x + a_0$ with real coefficients such that

$$\max_{-1 \leqslant x \leqslant 1} |x^n - a_{n-1}x^{n-1} - \cdots - a_0|$$

takes its smallest possible value. Clearly, this is equivalent to the following problem: among all polynomials $P(x) = x^n + a_{n-1}x^{n-1} + \cdots + a_0$ of degree n with leading coefficient 1, to find *the polynomial of least deviation from zero*, that is, a P with the smallest possible norm.

Since $1, x, \cdots, x^{n-1}$ is a Chebyshev system on $[-1, +1]$, Theorem 9 is applicable, and so we conclude that a polynomial P_n is a polynomial of least deviation from zero, if and only if (a) there exist $(n + 1)$ points of $[-1, +1]$ where P_n takes the values $\pm \| P_n \|$ with alternating signs, and (b) P_n has leading coefficient 1.

It is easy to find a P_n satisfying (a). For each x, $-1 \leqslant x \leqslant 1$, there is a unique t, $0 \leqslant t \leqslant \pi$, with $x = \cos t$. Consequently, $C_n(x) = \cos nt$, $n = 0, 1, \cdots$, are functions of x. We have $C_0(x) = 1$, $C_1(x) = x$. The C_n satisfy the recurrence relation

$$C_n(x) = 2xC_{n-1}(x) - C_{n-2}(x), \qquad n = 2, 3, \cdots, \tag{1}$$

which follows from

$$\cos nt + \cos (n - 2) t = 2 \cos t \cos (n - 1) t.$$

We compute easily:

$$C_2(x) = 2x^2 - 1, \qquad C_3(x) = 4x^3 - 3x, \qquad C_4(x) = 8x^4 - 8x^2 + 1, \cdots.$$

We see that C_n, $n = 0, 1, \cdots$, is an algebraic polynomial of degree n. The leading coefficient of C_n, $n = 1, 2, \cdots$, is 2^{n-1}. This follows from (1) by induction.

THEOREM 11. The polynomial of degree n with leading coefficient 1 of least deviation from zero on $[-1, +1]$ is $2^{-n+1}C_n(x)$. The degree of approximation of x^n on $[-1, +1]$ by polynomials of degree $n - 1$ is 2^{-n+1}.

Proof. We have $\| C_n \| = \max | \cos nt | = 1$. Also, $\cos nt = (-1)^k$ for $t = k\pi/n$, $k = 0, \cdots, n$; hence, $C_n(x_k) = (-1)^k$, where $x_k = \cos (k\pi/n)$, $k = 0, \cdots, n$. Thus, C_n satisfies (a). ∎

The polynomials C_n are called the *Chebyshev polynomials*. We note some of their remarkable properties.

1. The explicit expression for $C_n(x)$ in terms of x is

$$C_n(x) = \tfrac{1}{2}\{(x + \sqrt{x^2 - 1})^n + (x - \sqrt{x^2 - 1})^n\}. \tag{2}$$

Adding the two formulas

$$\cos nt \pm i \sin nt = (\cos t \pm i \sin t)^n,$$

we obtain

$$\cos nt = \tfrac{1}{2}\{(\cos t + i \sin t)^n + (\cos t - i \sin t)^n\}.$$

From this, (2) follows, for if $\cos t = x$, then $i \sin t$ is one of the two values of $\pm \sqrt{x^2 - 1}$.

2. The $C_n(x)$, $n = 0, 1, \cdots$, form a system of *orthogonal polynomials* on $[-1, +1]$ with weight-function $1/\sqrt{1 - x^2}$:

$$\int_{-1}^{+1} \frac{1}{\sqrt{1 - x^2}} C_n(x) C_m(x) \, dx = 0, \qquad n \neq m. \tag{3}$$

This follows from $\int_0^\pi \cos nt \cos mt \, dt = 0$, $n \neq m$.

3. The *zeros* of C_n can be easily found. Since $\cos nt = 0$ if $t = (2k-1)\pi/(2n)$,

$$x_k = \cos \frac{2k-1}{2n} \pi, \qquad k = 1, \cdots, n \tag{4}$$

are the n zeros of C_n. Thus, the zeros are always real, distinct, and lie in $[-1, +1]$. They are symmetric with respect to the point 0.

For large n, the points (4) are more dense around the points $-1, +1$ than toward the center of the interval $[-1, +1]$. They are often used with success as nodes of interpolation for continuous functions f on $[-1, +1]$. One of the reasons is perhaps the following. We use more values of $f(x)$ around the points $-1, +1$ than elsewhere, because we are less sure of $f(x)$ in the neighborhood of these points: $f(x)$ is not defined for $x < -1$ or $x > 1$.

8. Approximation of Some Complex Functions

We indicate a few cases, when the algebraic polynomial of best approximation can be found explicitly.

1. Let $f(z) = z^{n+1}$ on the disc $|z| \leqslant 1$. The algebraic polynomial of degree n of best approximation of f is zero, and therefore $E_n(f) = 1$.

To see this, it is sufficient to establish that $|f(z) - P_n(z)| < 1$, for all $|z| \leqslant 1$, is impossible for a polynomial P_n of degree n. For $z = e^{it}$, the inequality implies $|\operatorname{Re} f(e^{it}) - T_n(t)| < 1$, where $T_n(t) = \operatorname{Re} P_n(e^{it})$ is a trigonometric polynomial of degree n. Since $f(z) = (-1)^k$ for $z = e^{it_k}$, $t_k = k\pi/(n+1)$, $k = 0, \cdots, 2n + 1$, the polynomial T_n has at least $2n + 2$ changes of sign for $0 \leqslant t < 2\pi$, which is impossible.

2. We consider the function $f(z) = (z - \alpha)^{-1}$ on the disc A: $|z| \leqslant 1$; the number α, $|\alpha| > 1$ is a complex constant. Let

$$P_n(z) = \frac{1}{z - \alpha} - Cz^n \frac{\bar{\alpha}z - 1}{z - \alpha} = \frac{1 - Cz^n(\bar{\alpha}z - 1)}{z - \alpha}. \tag{1}$$

Clearly, P_n is a polynomial (of degree n) if and only if $1 - C\alpha^n(\bar{\alpha}\alpha - 1) = 0$. We put therefore

$$C = \frac{\alpha^{-n}}{|\alpha|^2 - 1}. \tag{2}$$

THEOREM 12 (Al'per [31]).   Let P_n and C be defined by (1) and (2). Then P_n is the polynomial of best approximation for f, and

$$E_n(f) = |C|. \tag{3}$$

Proof.   For $|z| = 1$,

$$\left| \frac{\bar{\alpha}z - 1}{z - \alpha} \right| = \left| \frac{(\alpha/z) - 1}{z - \alpha} \right| = 1,$$

and this expression does not exceed 1 for $|z| < 1$ because $(\bar{\alpha}z - 1)/(z - \alpha)$ is an analytic function in $|z| \leqslant 1$. Let $F = f - P_n$ and let A_0 be the subset of A where the function $|F(z)|$ attains its maximum $\|F\|$. We see that A_0 is exactly the circumference $|z| = 1$, and that $\|F\| = |C|$.

It remains to show that there does not exist a polynomial Q_n of degree n for which

$$|F(z) - Q_n(z)| < |C|, \qquad |z| \leqslant 1. \tag{4}$$

For this purpose we study the behavior of $F(z)$ for $z = e^{it}, 0 \leqslant t \leqslant 2\pi$. Since

$$\arg F(z) = \text{const} + n \arg z + \arg (z - \bar{\alpha}^{-1}) - \arg (z - \alpha),$$

this argument will increase by the amount $2\pi(n + 1)$ as t changes from 0 to 2π. There must exist $2n + 2$ points t_k, $0 \leqslant t_1 < t_2 < \cdots < t_{2n+2} < 2\pi$, for which the values of $\arg F(e^{it_k})$ are $2n + 2$ successive multiples of π. Then

$$F(e^{it_k}) = \mu(-1)^k |C|, \qquad k = 1, 2, \cdots,$$

with $\mu = +1$ or $\mu = -1$.

If (4) were true, then for the trigonometric polynomial $T_n(t) = \text{Re} \, Q_n(e^{it})$ we would have $|\text{Re} \, F(e^{it}) - T_n(t)| < |C|$. Thus T_n again would have at least $2n + 2$ changes of sign. This is impossible; hence, P_n is the polynomial of best approximation. Also, $E_n(f) = \|F\| = |C|$.   ∎

9. Notes

1.   For a fixed n, the algebraic polynomial P_n of best approximation of $f \in C[-1, +1]$ is uniquely determined by f. Thus, $P_n = P_n(f, x)$ is an operator on $C[-1, +1]$. One can prove that this operator is continuous. But P_n is not linear. Let, for example, $f(x) = \sin m\pi x$, $-1 \leqslant x \leqslant 1$, and let $g(x) = f(x)$ for $-1 \leqslant x \leqslant (m - 1)/m$, $g(x) = 0$ for $(m - 1)/m \leqslant x \leqslant 1$. The function f has $2m$ points of maximal deviation from 0 with alternating signs; the function g has $(2m - 1)$ such points. As long as $n + 2 \leqslant 2m - 1$, the polynomials of degree n of best approximation for f and g are both zero. But $f - g$ is better approximated by $\frac{1}{2}(-1)^{m-1}$ than by 0.

2.   The discussions of Sec. 3 leave open the question of which topological spaces A possess Chebyshev systems. The following theorem, due to Mairhuber [78] and Curtis [40], gives an answer.

THEOREM. A compact Hausdorff space A has nontrivial real Chebyshev systems if and only if A is homeomorphic to a subset of the circle K.

3. The importance of convexity in the questions of approximation was recognized by Haar [55] in 1918. In [105] Žuhovickiĭ gave an exposition, from this point of view, of the basic theorems of best approximation. In Secs. 3 and 6 we have followed the elegant paper of Rivlin and Shapiro [86].

4. Theorem 9, for the case $A = [a, b]$ and approximation by algebraic polynomials, was proved by Chebyshev in 1857. Our approach to this theorem somewhat obscures its elementary character; the reader should compare the proof of the original Chebyshev theorem in [10, p. 25]. A similar proof of the general form of Theorem 9 is based upon the condition 2(3) of Theorem 2; it uses the result of 4(4) and Problem 11.

5. We did not discuss problems of approximation of elements of an arbitrary Banach space. Important results in this direction are due to Phelps [85], Garkavi, and Singer. The forthcoming book of Butzer will contain a discussion of the results of Singer, which at this moment are available only in Rumanian.

PROBLEMS

1. Let a Banach space $X, f \in X, \phi_n \in X, n = 1, 2, \cdots$ be given. The degrees of approximation $E_n(f)$ of f by polynomials $P_n = a_1\phi_1 + \cdots + a_n\phi_n$ satisfy
$$E_n(f + g) \leqslant E_n(f) + E_n(g); \qquad E_n(af) = |a| E_n(f); \qquad E_n(f + P_n) = E_n(f);$$

$$E_1(f) \geqslant E_2(f) \geqslant \cdots .$$

2. The set of all polynomials P_n of best approximation to $f \in X$ is a closed convex subset of X.

3. If $f \in C[-1, +1]$ or $f \in C^*$ is even or odd, then the same is true for its polynomial of best approximation, P_n or T_n.

4. The polynomial $T_n(x) = a \cos nx + b \sin nx$ is of least derivation from zero among all trigonometric polynomials of degree n with prescribed coefficients of $\cos nx$ and $\sin nx$.

5. Prove that $e^{a_1x}, \cdots, e^{a_nx}$ and $x^{a_1}, \cdots, x^{a_n}$, where the a_i are distinct real numbers (and $a_1 = 1$, $a_i \geqslant 0$ for the second case) are two Chebyshev systems on $[0, 1]$.

6. If the function $f(x)$, $a \leqslant x \leqslant b$ has a continuous derivative $f^{(n)}(x)$, which does not vanish, then $1, x, \cdots, x^{n-1}, f(x)$ is a Chebyshev system on $[a, b]$.

7. Let

$$f(x) = \sum_{k=1}^{\infty} a_k \cos 3^k x, \quad \text{where} \qquad a_k > 0, \qquad \sum a_k < +\infty. \qquad (*)$$

The trigonometric polynomial of best approximation to $f(x)$ is the appropriate partial sum of the series (*), and the degree of approximation is

$$E_n^*(f) = \sum_{3^k > n} a_k .$$

8. There does not exist a polynomial

$$P_n(x) = a_0 x^n + \cdots + a_n , \qquad a_0 \neq 0, \quad n \geqslant 1,$$

with integral coefficients that satisfies $|P_n(x)| < 2$ on an interval $[a, b]$ of length $b - a \geqslant 4$.

9. Using Problem 8, show that a continuous function on $[a, b]$, $b - a \geqslant 4$, which is not a polynomial, cannot be approximable by polynomials with integral coefficients.

10. If P_n and Q_m are algebraic polynomials of best approximation to f and g, respectively, on $[0, 1]$, then $P_n(x) + Q_m(y)$ is a polynomial of best approximation to $f(x) + g(y)$ $\left(\text{among the admissible polynomials } \sum_{k=0}^{n} \sum_{l=0}^{m} a_{kl} x^k y^l \right)$ (Newman and Shapiro). [*Hint:* Construct an extremal signature that satisfies 3(9) for $f(x) + g(y) - P_n(x) - Q_m(y)$.]

11. For a real Chebyshev system on $A = [a, b]$ or $A = K$, there is a polynomial $P(x)$ for which $P(x) > 0$, $x \in A$.

[3]

Properties of Polynomials
and Moduli of Continuity

1. Interpolation

Let Φ: ϕ_1, $\cdots$, ϕ_n be a system of real-valued functions on a set A; let x_1, $\cdots$, x_n be given distinct points of A; and let c_1, $\cdots$, c_n be given real numbers. The polynomial $P(x) = \sum\limits_{i=1}^{n} a_i \phi_i(x)$ is said to *interpolate* the values c_k if $P(x_k) = c_k$, $k = 1$, $\cdots$, n. Usually, the c_k are values at the points x_k of some given function $f(x)$; then P is said to interpolate f. Assume that we can find polynomials $l_k(x)$, $k = 1$, $\cdots$, n such that

$$l_k(x_k) = 1, \qquad l_k(x_i) = 0, \qquad i \neq k, \qquad i, k = 1, 2, \cdots, n. \qquad (1)$$

Then an interpolating polynomial P can be written in the form

$$P(x) = \sum_{i=1}^{n} c_i l_i(x); \qquad (2)$$

it is clear that $P(x_k) = c_k$ for each k. If Φ is a Chebyshev system, the polynomial $P(x)$ is unique. In this case, we can take

$$l_k(x) = \frac{D(x_1, \cdots, x_{k-1}, x, x_{k+1}, \cdots, x_n)}{D(x_1, \cdots, x_{k-1}, x_k, x_{k+1}, \cdots, x_n)}, \qquad (3)$$

where $D(x_1, \cdots, x_n)$ is the determinant 2.4(3). Special cases (see (4) and (6) below) may be handled by means of (3), or directly.

1. Algebraic Polynomials on $[a, b]$. Let x_0, $\cdots$, x_n be $(n + 1)$ different points of the interval $[a, b]$, c_0, $\cdots$, c_n given numbers. We look for an algebraic polynomial $P_n(x)$ of degree n that assumes values c_k at the points x_k. We can select

$$l_k(x) = \frac{(x - x_0) \cdots (x - x_{k-1}) (x - x_{k+1}) \cdots (x - x_n)}{(x_k - x_0) \cdots (x_k - x_{k-1}) (x_k - x_{k+1}) \cdots (x_k - x_n)}, \qquad k = 0, \cdots, n. \qquad (4)$$

If we put $\Omega(x) = (x - x_0) \cdots (x - x_n)$, we can write this formula in the form

$$l_k(x) = \frac{\Omega(x)}{(x - x_k)\, \Omega'(x_k)}.$$

Thus, the interpolating polynomial is given by

$$P(x) = \sum_{k=0}^{n} \frac{c_k \Omega(x)}{(x - x_k)\, \Omega'(x_k)}. \tag{5}$$

This is *Lagrange's interpolation formula*. If fewer than $n + 1$ different points are given, the interpolating polynomial is not unique. We may change the situation by prescribing not only the values of $P_n(x)$ at $x = x_k$, but also values of some of the derivatives $P_n^{(j)}(x)$ at $x = x_k$, $j = 1, 2, \cdots, m_k$. This again determines $P_n(x)$ uniquely if the total number of given values is equal to $n + 1$, the number of coefficients of P_n. We do not give the corresponding formula, which is called *the interpolation formula of Hermite*. Its extreme cases are Lagrange's formula (5), if all $n + 1$ points are different, and Taylor's formula, which gives $P_n(x)$ with prescribed $P_n(x_0), P_n'(x_0), \cdots, P_n^{(n)}(x_0)$ when all $n + 1$ interpolation points coincide.

One might expect that if $P_n(x)$ is the interpolating polynomial (5) of a continuous function $f(x)$, then we must have $P_n(x) \to f(x)$ if $n \to \infty$ and if the interpolation points become dense in $[a, b]$. But this is, in general, not so. For example, Bernstein showed that if the interpolation points are equidistant in $[-1, +1]$ with $x_0 = -1$, $x_n = +1$, then for $f(x) = |x|$, the interpolating polynomials $P_n(x)$ diverge at each point of $[-1, +1]$ except for $x = 0$, $x = -1$, and $x = +1$! (See [10, p. 375].)

2. Trigonometric Polynomials. Let $x_0, \cdots, x_{2n}$ be $2n + 1$ distinct points of the circle K. Then the trigonometric polynomial $l_k(x)$ of degree n, which satisfies $l_k(x_k) = 1$, $l_k(x_i) = 0$, $i \neq k$, is given by

$$l_k(x) = \frac{\sin \dfrac{x - x_0}{2} \cdots \sin \dfrac{x - x_{k-1}}{2} \sin \dfrac{x - x_{k+1}}{2} \cdots \sin \dfrac{x - x_{2n}}{2}}{\sin \dfrac{x_k - x_0}{2} \cdots \sin \dfrac{x_k - x_{k-1}}{2} \sin \dfrac{x_k - x_{k+1}}{2} \cdots \sin \dfrac{x_k - x_{2n}}{2}}; \tag{6}$$

one should notice that products of the type $\sin((x - x_0)/2) \sin((x - x_1)/2)$ can be written in the form $a + b \cos x + c \sin x$, so that $l_k(x)$ is indeed a polynomial of degree n.

Returning to the case of a real Chebyshev system of n functions on $A = [a, b]$ or $A = K$, we remark that the polynomial of best approximation $P(x)$ of a continuous function $f(x)$ is also an interpolating polynomial of $f(x)$ at some n points of A. In fact, by Chebyshev's theorem (Theorem 9, Chapter 2), the difference $f(x) - P(x)$ takes values of alternating signs at some $n + 1$ points of A; by continuity, there are at least n points where $f(x) - P(x)$ vanishes. This simple remark allows the following application:

THEOREM 1 (Bernstein [2, vol. I, p. 63]).   Let $f(x)$, $g(x)$ be two functions with continuous derivatives of order $n + 1$ on the interval $[a, b]$, and let

$$|f^{(n+1)}(x)| \leqslant g^{(n+1)}(x), \qquad a \leqslant x \leqslant b. \tag{7}$$

Then the degrees of approximation by algebraic polynomials satisfy the inequality

$$E_n(f) \leqslant E_n(g). \tag{8}$$

Proof. We assume first that $g^{(n+1)}(x)$ does not vanish for $a < x < b$. We begin by taking $n + 1$ arbitrary different points x_0, $\cdots$, x_n of $[a, b]$. Let P_n and Q_n be the interpolating polynomials of f and g at these points and put $R = f - P_n$, $S = g - Q_n$. Let x be a fixed point of $[a, b]$ different from each point x_k. The function of y,

$$\Phi(y) = R(x) S(y) - S(x) R(y),$$

has $n + 2$ different zeros, $y = x, x_0, \cdots, x_n$. Hence, by Rolle's theorem, its $(n + 1)$st derivative vanishes at some point ξ of (a, b). But $R^{(n+1)} = f^{(n+1)}$, $S^{(n+1)} = g^{(n+1)}$, and we obtain

$$R(x) g^{(n+1)}(\xi) = S(x) f^{(n+1)}(\xi).$$

Using (7), we have

$$|f(x) - P_n(x)| \leqslant |g(x) - Q_n(x)|. \tag{9}$$

Relation (9) is true also for $x = x_0, \cdots, x_n$; hence, it holds for all x.

We shall use (9) by taking for $Q_n(x)$ the polynomial of best approximation to $g(x)$, and for $x_0, \cdots, x_n$ we shall take a set of $n + 1$ points where $Q_n(x)$ is equal to $g(x)$. Then (9) shows that

$$|f(x) - P_n(x)| \leqslant E_n(g)$$

for the corresponding interpolating polynomial P_n of f. Hence, (8) must hold.

If $g^{(n+1)}$ has zeros, we add to it a small function with nonvanishing derivative; for example, ϵh, $h(x) = (x - a)^{n+1}$, $\epsilon > 0$. Then

$$E_n(f) \leqslant E_n(g + \epsilon h) \leqslant E_n(g) + \epsilon \| h \|,$$

and for $\epsilon \to 0$, we obtain (8).   ∎

COROLLARY.   Let $|f^{(n+1)}(x)| \leqslant M$ in $[-1, +1]$. Then $E_n(f) \leqslant M2^{-n}/(n+1)!$ For the proof, we take $g(x) = Mx^{n+1}/(n+1)!$ and note that by Theorem 11, Chapter 2, the degree of approximation of x^{n+1} by polynomials of degree n is 2^{-n}.

2. Inequalities of Bernstein

Many theorems of the theory of approximation depend upon the fact that a polynomial of degree n cannot change too rapidly; in other words, its derivative cannot be too large. In the following theorems, we estimate the magnitude of the derivative, as compared to the polynomial itself.

We first discuss trigonometric polynomials. If $T_n(x) = \cos nx$ or $T_n(x) = \sin nx$, then $\| T_n \| = 1$, and the derivative T_n' has the norm equal to n. This is the extreme case.

THEOREM 2 (Bernstein [2, vol. I, p. 26]). If a trigonometric polynomial $T_n(x)$ satisfies on K the inequality $| T_n(x) | \leqslant M$, then

$$| T_n'(x) | \leqslant nM, \qquad x \in K. \tag{1}$$

Proof. If the theorem is not true, then there exists a polynomial T_n with $\| T_n' \| = nL, L > \| T_n \|$. At some point $x_0 \in K$, $| T_n'(x_0) | = nL$, and we can assume that $T_n'(x_0) = nL$. Since T_n' has a maximum at x_0, $T_n''(x_0) = 0$.

Now consider the trigonometric polynomial of degree n:

$$S_n(x) = L \sin n(x - x_0) - T_n(x). \tag{2}$$

At the $2n$ points of K, $x = x_0 + (2k + 1) \pi(2n)^{-1}$, where $\sin n(x - x_0)$ takes the values ± 1, the polynomial S_n takes values of alternating signs. Between any two of these points, S_n has a zero; hence, S_n has $2n$ different zeros on K. By Rolle's theorem,

$$S_n'(x) = nL \cos n(x - x_0) - T_n'(x)$$

also has $2n$ different zeros. One of these zeros is x_0, since

$$S_n'(x_0) = nL - T_n'(x_0) = 0.$$

Also,

$$S_n''(x) = - n^2 L \sin n(x - x_0) - T_n''(x) \tag{3}$$

vanishes at $x = x_0$. Moreover, S_n'' has, again by Rolle's theorem, $2n$ zeros between the zeros of S_n'. Thus, S_n'' has at least $2n + 1$ zeros; hence (by Chapter 2, Sec. 4), S_n'' is identically zero. Then S_n is constant; this is a contradiction, since we have found that S_n changes sign. Thus, (1) must hold. ∎

The corresponding statement about algebraic polynomials $P_n(x)$ of degree n is the following

THEOREM 3. Let $| P_n(x) | \leqslant M$ for $- 1 \leqslant x \leqslant 1$. Then

$$| P_n'(x) | \leqslant \frac{Mn}{\sqrt{1 - x^2}}, \qquad - 1 < x < 1. \tag{4}$$

Proof. This can be obtained from Theorem 2 by the ,,standard" substitution $x = \cos t$. Since $T_n(t) = P_n(\cos t)$ is a trigonometric polynomial of degree n, and since $\mid T_n(t) \mid \leqslant M$, we have $\mid T_n{}'(t) \mid \leqslant Mn$. But

$$T_n'(t) = \pm P_n'(x)\sqrt{1 - x^2}. \qquad \blacksquare$$

From Theorems 2 and 3 we can also derive (by iteration) estimates of derivatives of arbitrary orders. For example, for trigonometric polynomials, if $\| T_n \| \leqslant M$, then

$$\| T_n^{(p)} \| \leqslant Mn^p, \qquad p = 1, 2, \cdots. \qquad (5)$$

Theorem 2 remains true for trigonometric polynomials T_n *with complex coefficients.* We select a real α in such a way that $e^{i\alpha}T_n'$ attains the value $\| T_n' \|$; say, for $x = x_0$. Now $S_n(x) = \operatorname{Re}\{e^{i\alpha}T_n(x)\}$ is a polynomial with real coefficients, for which $\| S_n \| \leqslant \| T_n \|$ and $S_n'(x) = \operatorname{Re}\{e^{i\alpha}T_n'(x)\}$. Hence,

$$\| T_n' \| = e^{i\alpha}T_n'(x_0) = S_n'(x_0) \leqslant n \| T_n \|.$$

From this we obtain

THEOREM 4.   For polynomials $P_n(z) = \sum_0^n a_k z^k$ with complex coefficients on the disc $\mid z \mid \leqslant 1$, we put

$$\| P_n \| = \max_{|z| \leqslant 1} \mid P_n(z) \mid.$$

Then

$$\| P_n' \| \leqslant n \| P_n \|. \qquad (6)$$

Proof. We consider P_n and P_n' on the circumference $\mid z \mid = 1$; that is, for $z = e^{ix}$, $x \in K$. Then $P_n(z) = \sum_0^n a_k e^{ikx} = T_n(x)$ is a trigonometric polynomial with complex coefficients, and $P_n'(z) = \sum_1^n a_k k e^{i(k-1)x} = -ie^{-ix}T_n'(x)$. Therefore, (6) follows from (1), and the maximum modulus principle. $\blacksquare$

3. The Inequality of Markov

The inequality 2(4) is not always the best possible: If x approaches the end points -1, $+1$ of the interval, the upper bound becomes arbitrarily large. One can, however, give a constant upper bound.

THEOREM 5 (A. A. Markov [79]).   If $\mid P_n(x) \mid \leqslant M$ on $[-1, +1]$, then

$$\mid P_n'(x) \mid \leqslant Mn^2, \qquad -1 \leqslant x \leqslant 1. \qquad (1)$$

It is clear that (1) cannot be improved. If P_n is a Chebyshev polynomial

$C_n(x) = \cos nt$, $x = \cos t$, $0 \leqslant t \leqslant \pi$, then $\| C_n \| = 1$ and $C_n'(x) = n \sin nt / \sin t$ is equal to n^2 for $x = 1$ (or $t = 0$). We see also that

$$| C_n'(x) | \leqslant n^2, \qquad -1 \leqslant x \leqslant 1. \tag{2}$$

This follows from the inequality $| \sin nt | \leqslant n \, | \sin t |$, which can be easily proved by induction in n. Markov's theorem follows immediately from Theorem 3 and the following.

THEOREM 6 (Schur).   If S_{n-1} is an algebraic polynomial of degree $n - 1$ that satisfies $| S_{n-1}(x) | \leqslant L(1 - x^2)^{-1/2}$ for $-1 < x < 1$, then $| S_{n-1}(x) | \leqslant Ln$ on this interval.

Proof.   Let x_k, $-1 < x_n < \cdots < x_1 < 1$ be the zeros of the Chebyshev polynomial C_n [see 2.7(4)]. We consider two cases.

(a)   First let $| x | \leqslant x_1$ or $x_n \leqslant x \leqslant x_1$. Then

$$(1 - x^2)^{-1/2} \leqslant (1 - x_1^2)^{-1/2} = \left(1 - \cos^2 \frac{\pi}{2n}\right)^{-1/2} = \left(\sin \frac{\pi}{2n}\right)^{-1} \leqslant n$$

because of the inequality $\sin x \geqslant (2/\pi) x$, which is valid for $0 \leqslant x \leqslant \pi/2$ (this follows from the fact that the curve $y = \sin x$ is convex for $0 \leqslant x \leqslant \pi/2$ and is therefore above its chord $y = (2/\pi) x$).

(b)   For other values of x, we interpolate S_{n-1} at the zeros x_k of the Chebyshev polynomial $C_n(x)$ according to 1(5) and obtain

$$S_{n-1}(x) = \sum_{k=1}^{n} S_{n-1}(x_k) \frac{C_n(x)}{(x - x_k) \, C_n'(x_k)}. \tag{3}$$

With $x = \cos t$, the zeros x_k of C_n correspond to the values $t_k = (2k - 1)\pi(2n)^{-1}$, $k = 1, \cdots, n$ of t; hence,

$$| C_n'(x_k) | = \left| n \, \frac{\sin nt_k}{\sqrt{1 - x_k^2}} \right| = \frac{n}{\sqrt{1 - x_k^2}}$$

Replacing $S_{n-1}(x_k)$ in (3) by its upper bound according to the assumptions, we obtain

$$| S_{n-1}(x) | \leqslant Ln^{-1} \sum_{k=1}^{n} \left| \frac{C_n(x)}{x - x_k} \right|, \qquad -1 \leqslant x \leqslant 1.$$

But $C_n(x) = \text{const} \, (x - x_1) \cdots (x - x_n)$, so that for each x with $x \leqslant x_n$ or $x \geqslant x_1$, all quotients $C_n(x)/(x - x_k)$ have the same sign. By (2),

$$| S_{n-1}(x) | \leqslant Ln^{-1} \left| \sum_{k=1}^{n} \frac{C_n(x)}{x - x_k} \right| = Ln^{-1} | C_n'(x) | \leqslant Ln.$$

Together, (a) and (b) establish the theorem.

Later (Chapter 5, Sec. 3; Chapter 7, Sec. 5) we shall establish some further inequalities concerning the derivatives of polynomials.

4. Growth of Polynomials in the Complex Plane

If a polynomial P_n satisfies $|P_n(x)| \leqslant 1$ on $[-1, +1]$, what can be said about its behavior in the complex plane? To answer this question, we consider the ellipses E_ρ, defined as follows: The transformation

$$z = \frac{1}{2}\left(w + \frac{1}{w}\right) \tag{1}$$

can be written in the real form

$$x = \tfrac{1}{2}(\rho + \rho^{-1})\cos\phi$$
$$y = \tfrac{1}{2}(\rho - \rho^{-1})\sin\phi, \tag{2}$$

where $z = x + iy$, $w = \rho e^{i\phi}$. The curve E_ρ with constant $\rho > 0$ is for $\rho \neq 1$, the ellipse $x^2/a^2 + y^2/b^2 = 1$ with the semiaxes $a = \tfrac{1}{2}(\rho + \rho^{-1})$ and $b = \tfrac{1}{2}|\rho - \rho^{-1}|$, and the foci ± 1. The ellipse $E_{\rho^{-1}}$ coincides with E_ρ. For $\rho = 1$, E_ρ is the interval $[-1, +1]$ covered twice. For each point z not on $[-1, +1]$, there is exactly one ellipse E_ρ, $\rho > 1$ passing through z; ρ is found from the equation $\rho + \rho^{-1} = |z - 1| + |z + 1|$. It follows that the transformation (1) maps the region $|w| > 1$ in a one-to-one way onto the complement of the interval $[-1, +1]$ in the z plane.

THEOREM 7 (Bernstein 2, vol. I, p. 21). If a polynomial P_n with complex coefficients satisfies $|P_n(x)| \leqslant M$ for $-1 \leqslant x \leqslant 1$, then

$$|P_n(z)| \leqslant M\rho^n, \qquad z \in E_\rho, \qquad \rho > 1. \tag{3}$$

Proof. Assume that $\rho > 1$ is given. We construct the polynomial of degree $2n$,

$$Q(w) = w^n P_n\left(\frac{w + w^{-1}}{2}\right),$$

and apply to it the maximum modulus principle of analytic function:

$$\max_{|w|=\rho^{-1}} |Q(w)| \leqslant \max_{|w|=1} |Q(w)| = \max_{|w|=1}\left|P_n\left(\frac{w + w^{-1}}{2}\right)\right|$$
$$= \max_{-1 \leqslant x \leqslant 1} |P_n(x)| \leqslant M. \tag{4}$$

If w is on the circle $|w| = \rho^{-1}$, $z = \tfrac{1}{2}(w + w^{-1})$ is on the ellipse E_ρ, so that (4) leads directly to (3). ∎

As examples of application of Theorem 7, we have for polynomials on the real axis the following:

1. If $P_n(x)$ is a polynomial of degree n, $0 < a < 1$, and $| P_n(x) | \leqslant M$ for $| x | \leqslant a$, then

$$| P_n(x) | \leqslant M \left(\frac{1 + \sqrt{1 - a^2}}{a} \right)^n, \qquad | x | \leqslant 1. \qquad (5)$$

2. If $| P_n(x) | \leqslant M$ for $| x | \leqslant 1 - n^{-2}$, $n > 1$, then

$$| P_n(x) | \leqslant CM, \qquad | x | \leqslant 1, \qquad (6)$$

with some absolute constant C.

The first statement follows from Theorem 7, applied to $Q_n(z) = P_n(az)$, and the second is a special case of the first statement with $a = 1 - n^{-2}$.

5. Moduli of Continuity

To measure the continuity of a function $f \in C[A]$, if A is the set $[a, b]$ or K, we proceed as follows: We consider the *first difference* with step t,

$$\Delta_t f(x) = f(x + t) - f(x),$$

of the function f and put

$$\omega(f, h) = \omega(h) = \max_{\substack{x, t \\ |t| \leqslant h}} | f(x + t) - f(x) | . \qquad (1)$$

The function $\omega(h)$, called the *modulus of continuity* of f, is defined for $0 \leqslant h \leqslant l$, where $l = b - a$ if $A = [a, b]$, and $l = \pi$ if $A = K$.

A modulus of continuity $\omega(h)$ has the following properties: (a) $\omega(h) \to 0$ for $h \to 0$; (b) $\omega(h)$ is positive and increasing; (c) ω is subadditive:

$$\omega(h_1 + h_2) \leqslant \omega(h_1) + \omega(h_2). \qquad (2)$$

For if $| t | \leqslant h_1 + h_2$, then t can be put into form $t = t_1 + t_2$, $| t_1 | \leqslant h_1$, $| t_2 | \leqslant h_2$, and (2) follows at once from

$$| f(x + t) - f(x) | \leqslant | f(x + t_1 + t_2) - f(x + t_2) | + | f(x + t_2) - f(x) | .$$

Moreover, we have $| \omega(h_1 \pm h_2) - \omega(h_1) | \leqslant \omega(h_2)$. Therefore, (a), (b), and (c) imply the property (d): $\omega(h)$ is continuous.

Conditions (a), (b), (c) are also *sufficient* for a given function $\omega(h)$, $0 \leqslant h \leqslant l$, to be the modulus of continuity of some function $f \in C[A]$. In fact, the modulus of continuity of $f(x) = \omega(x)$, $0 \leqslant x \leqslant l$, is $\omega(x)$ itself, and this also applies to the functions $f(x) = \omega(x - a)$, $a \leqslant x \leqslant b$, and $f(x) = \omega(| x |)$, $-\pi \leqslant x \leqslant \pi$.

It follows from (2) by induction that

$$\omega(h_1 + \cdots + h_n) \leqslant \omega(h_1) + \cdots + \omega(h_n).$$

For $h = h_1 = \cdots = h_n$, we obtain

$$\omega(nh) \leqslant n\omega(h). \tag{3}$$

A similar inequality holds for a nonintegral factor λ:

$$\omega(\lambda h) \leqslant (\lambda + 1)\,\omega(h), \qquad \lambda \geqslant 0. \tag{4}$$

In fact, taking an integer n for which $n \leqslant \lambda < n + 1$, we see that

$$\omega(\lambda h) \leqslant \omega((n + 1)\,h) \leqslant (n + 1)\,\omega(h) \leqslant (\lambda + 1)\,\omega(h).$$

Let F be a product, $F(x) = f_1(x)\,f_2(x)$. From the identity

$$\Delta_t F(x) = f_1(x + t)\,\Delta_t f_2(x) + f_2(x)\,\Delta_t f_1(x),$$

we obtain

$$\omega(F, h) \leqslant \|f_2\|\,\omega(f_1, h) + \|f_1\|\,\omega(f_2, h).$$

The corresponding inequality for a product of s functions, f_i, has the form

$$\omega(f_1 \cdots f_s, h) \leqslant \sum_{i=1}^{s} \|f_1\| \cdots \|f_{i-1}\|\,\omega(f_i, h)\,\|f_{i+1}\| \cdots \|f_s\|. \tag{5}$$

The modulus of continuity cannot be too small: If $\omega(h)/h \to 0$ for $h \to 0$, then the function f is a constant. Indeed, we have in this case $|\,[f(x + t) - f(x)]/t\,| \to 0$ as $t \to 0$; therefore, $f'(x)$ exists and is equal to zero for all $x \in A$. As a function of f, the modulus of continuity has the properties of a seminorm:

$$\omega(af, h) = |a|\,\omega(f, h); \qquad \omega(f_1 + f_2, h) \leqslant \omega(f_1, h) + \omega(f_2, h). \tag{6}$$

A function $\omega(x)$ is *concave* on $[a, b]$ if it satisfies the inequality $\alpha f(x) + \beta f(y) \leqslant f(\alpha x + \beta y)$ for arbitrary $x, y \in [a, b]$ and $\alpha \geqslant 0$, $\beta \geqslant 0$, $\alpha + \beta = 1$. A concave function $f(x)$ on $[0, l]$, which satisfies $f(0) = 0$, has the property that $f(x)/x$ decreases, for if $x < y$, then

$$\frac{x}{y} f(y) = \frac{y - x}{y} f(0) + \frac{x}{y} f(y) \leqslant f(x).$$

A continuous, increasing function $\omega(x)$ on $[0, l]$, which satisfies $\omega(0) = 0$, is a modulus of continuity if it is concave (or, more generally, if $\omega(x)/x$ is decreasing). It is necessary only to show that ω satisfies (2). This is obtained by multiplying the inequalities

$$\frac{\omega(h_1 + h_2)}{h_1 + h_2} \leqslant \frac{\omega(h_1)}{h_1} \quad \text{and} \quad \frac{\omega(h_1 + h_2)}{h_1 + h_2} \leqslant \frac{\omega(h_2)}{h_2}$$

with h_1 and h_2, respectively, and adding.

We consider some examples, which will be useful later. Let $I_i = (\alpha_i, \alpha_i + 2\delta)$, $\delta > 0$, be finitely many disjoint intervals contained in $[a, b]$. Let $F(x)$ be zero outside the I_i, and let

$$F(x) = \begin{cases} c_i\omega(x - \alpha_i), & \alpha_i \leqslant x \leqslant \alpha_i + \delta, \\ c_i\omega(\alpha_i + 2\delta - x), & \alpha_i + \delta \leqslant x \leqslant \alpha_i + 2\delta. \end{cases} \tag{7}$$

If there is only one interval I_1 with $c_1 = c > 0$, and if $h \leqslant \delta$, then the maximum of $|F(x + t) - F(x)|$ for $|t| \leqslant h$ is attained when $x = \alpha_1$, $t = h$, and is equal to $c\omega(h)$. It is $c\omega(\delta)$ if $h \geqslant \delta$. Hence,

$$\omega(F, h) = \begin{cases} c\omega(h), & 0 \leqslant h \leqslant \delta, \\ c\omega(\delta), & h \geqslant \delta. \end{cases} \tag{8}$$

This remains true for several I_i if all $c_i > 0$ and $c = \max c_i$. If the c_i change sign, let $c = \max |c_i|$. Then F is the difference of two functions of type (7) with positive c_i, and we have (8) with c replaced by $2c$.

However, if ω is concave, a better estimate holds:

$$\omega(F, h) \leqslant \begin{cases} 2c\omega(\tfrac{1}{2} h), & 0 \leqslant h \leqslant 2\delta, \\ 2c\omega(\delta), & h \geqslant 2\delta. \end{cases} \tag{9}$$

In fact, the largest possible value of $|F(x + h) - F(x)|$, $0 < h \leqslant 2\delta$, happens when $x \in I_{i-1}$, $(x + h) \in I_i$, with the intervals I_{i-1}, I_i having the common end point α_i, and when $c_{i-1} = -c_i = \pm c$. If $x = \alpha_i - h_1$, $0 \leqslant h_1 \leqslant h$, then

$$|F(x + h) - F(x)| \leqslant c\omega(h_1) + c\omega(h - h_1) \leqslant 2c\omega\left(\tfrac{1}{2} h\right).$$

A modulus of continuity need not be concave; however, we have

THEOREM 8. For each modulus of continuity $\omega(x)$, $0 \leqslant x \leqslant l$, there is a concave modulus of continuity $\omega_1(x)$ with the property .

$$\omega(x) \leqslant \omega_1(x) \leqslant 2\omega(x), \qquad 0 \leqslant x \leqslant l. \tag{10}$$

Proof. Let B denote the subset of the plane consisting of all points (x, y) for which $0 \leqslant x \leqslant l$, $0 \leqslant y \leqslant \omega(x)$. The set B is compact, and the same is true for its convex hull B_c (see Chapter 2, Sec. 3). The intersection of B_c with a line $x = x_0$ $(0 \leqslant x_0 \leqslant l)$ is a closed segment of the form $[0, y_0]$. We put $y_0 = \omega_1(x_0)$.

The function ω_1 is concave on $[0, l]$. In fact, let $0 \leqslant x_1 < x_2 \leqslant l$. Since the points $(x_1, \omega_1(x_1))$ and $(x_2, \omega_1(x_2))$ belong to B_c, the same is true for all points $(\alpha x_1 + \beta x_2, \alpha\omega_1(x_1) + \beta\omega_1(x_2))$, $\alpha \geqslant 0$, $\beta \geqslant 0$, $\alpha + \beta = 1$. Therefore,

$$\alpha\omega_1(x_1) + \beta\omega_1(x_2) \leqslant \omega_1(\alpha x_1 + \beta x_2). \tag{11}$$

We define, for $0 \leqslant x_1 < x_2 \leqslant l$,

$$\omega(x_1, x_2; x) = \begin{cases} \dfrac{x_2 - x}{x_2 - x_1}\,\omega(x_1) + \dfrac{x - x_1}{x_2 - x_1}\,\omega(x_2), & x_1 \leqslant x \leqslant x_2, \\ \\ \omega(x) \qquad \text{outside} \quad [x_1, x_2]. \end{cases} \qquad (12)$$

Then

$$\omega_1(x) = \max_{x_1 \leqslant x \leqslant x_2} \omega(x_1, x_2; x). \qquad (13)$$

The inequality $\omega_1(x) \geqslant \omega(x_1, x_2; x)$ (for $x_1 \leqslant x \leqslant x_2$) follows from the definition of ω_1. On the other hand, by Lemma 2, Chapter 2, for each x, $0 \leqslant x \leqslant l$, the point $(x, \omega_1(x))$ is contained in a triangle $\varDelta$, whose three corners belong to the set B.

We can exclude the case in which $(x, \omega_1(x))$ belongs to a vertical side of $\varDelta$. Then $\omega_1(x) \leqslant l(x)$, where $y = l(x)$ is the equation of one of the three sides of $\varDelta$. But $l(x) \leqslant \omega(x_1, x_2; x)$ for properly chosen x_1, x_2.

The function ω_1 satisfies (10). The first inequality is obvious. On the other hand, let $x_1 \leqslant x \leqslant x_2$, $x_1 < x_2$. We have $\omega(x_1) \leqslant \omega(x)$ and

$$\omega(x_2) \leqslant (x_2 + x)\,x^{-1}\omega(x)$$

by (4). Therefore,

$$\omega(x_1, x_2; x) \leqslant \frac{\omega(x)}{x_2 - x_1}\left\{(x_2 - x) + \frac{x_2 + x}{x}\,(x - x_1)\right\}.$$

The expression enclosed in braces is equal to

$$x_2 - x_1 + \frac{x_2}{x}\,(x - x_1) \leqslant 2(x_2 - x_1);$$

hence, $\omega(x_1, x_2; x) \leqslant 2\omega(x)$.

It remains to add that ω_1 is increasing, since the functions $\omega(x_1, x_2; x)$ in (13) increase. Also, $\omega_1(0) = 0$, and ω_1 is continuous at $x = 0$ [by (10)] and at $x = l$ [by (10)], and since $\omega_1(l) = \omega(l)$, it is continuous at the interior points of $[0, l]$ (this is the property of each concave increasing function). Thus, ω_1 is a modulus of continuity. ▮

We also consider functions $f(x_1, \cdots, x_s)$ of s variables on A, where A is either an s-dimensional parallelepiped (that is, the product of s intervals $a_k \leqslant x_k \leqslant b_k$, $k = 1, \cdots, s$) or an s-dimensional torus $K^{(s)}$, the product of s circles K. We define the modulus of continuity $\omega(f, h)$ as the maximum of

$$|f(y_1, \cdots, y_s) - f(x_1, \cdots, x_s)| \qquad \text{for} \qquad |y_k - x_k| \leqslant h, \quad k = 1, \cdots, s. \qquad (14)$$

Sometimes the partial moduli of continuity $\omega^{(k)}(f, h)$ of f are useful. The kth modulus among them is the maximum of (14) when the increment is only with respect to the kth coordinate: $x_i = y_i$, $i \neq k$, $|y_k - x_k| \leqslant h$.

6. Moduli of Smoothness

Moduli of continuity $\omega(f, h)$ will play an important role in Chapter 4 and 5 in connection with the estimation of the degree of approximation of a function f by trigonometric or algebraic polynomials. Moduli of continuity of higher orders or *moduli of smoothness* can also be used for this purpose. They are connected with *differences of order* $r = 1, 2, \cdots$, with step t, $\Delta_t^r f(r)$ of the function f, which are defined by induction. If $f(x)$ is defined on $A = [a, b]$ or $A = K$, then $\Delta_t^1 f(x) = \Delta_t f(x)$; $\Delta_t^r f(x) = \Delta_t \Delta_t^{r-1} f(x)$. One proves by induction:

$$\Delta_t^r f(x) = \sum_{k=0}^{r} (-1)^{r-k} \binom{r}{k} f(x + kt). \tag{1}$$

Another useful formula is valid for functions f, which have absolutely continuous derivative $f^{(r-1)}$ (in this case, $f^{(r)}$ exists almost everywhere on A, and $f^{(r-1)}$ is an indefinite integral of $f^{(r)}$):

$$\Delta_t^r f(x) = \int_0^t \cdots \int_0^t f^{(r)}(x + y_1 + \cdots + y_r)\, dy_1 \cdots dy_r. \tag{2}$$

This formula also is obvious for $r = 1$, and is proved by induction on r in the general case.

If $l = b - a$ (for $A = [a, b]$) or $l = \pi$ (for $A = K$), the moduli of smoothness of $f \in C[A]$ are defined by

$$\omega_r(f, h) = \max_{\substack{x, t \\ |t| \leqslant h}} |\Delta_t^r f(x)|, \qquad r = 1, 2, \cdots, \qquad 0 \leqslant h \leqslant l/r; \tag{3}$$

$\omega_1(f, h)$ is the modulus of continuity $\omega(f, h)$. As a function of h, $\omega_r(f, h)$ is continuous, increasing, and satisfies $\omega_r(f, 0) = 0$. If formula (2) applies and $f^{(r)}$ is bounded, we can estimate the difference (2):

$$|\Delta_t^r f(x)| \leqslant t^r \sup_x |f^{(r)}(x)|.$$

This gives

$$\omega_r(f, h) \leqslant h^r \sup_x |f^{(r)}(x)|. \tag{4}$$

Let $f^{(r)}$ be continuous. From (2) we see that $\Delta_t^{r+s} f(x)$, $s = 1, 2, \cdots$, is an r-tuple integral of $\Delta_t^s f^{(r)}(x + y_1 + \cdots + y_r)$. Applying to this integral the mean-value theorem, and then taking maxima for $x \in A$, $|t| \leqslant h$ on both sides, we obtain

$$\omega_{r+s}(f, h) \leqslant h^r \omega_s(f^{(r)}, h). \tag{5}$$

We also need a formula similar to 5(3), which follows from the relation

$$\Delta_{nt}^r f(x) = \sum_{k_1=0}^{n-1} \cdots \sum_{k_r=0}^{n-1} \Delta_t^r f(x + k_1 t + \cdots + k_r t). \tag{6}$$

If $r = 1$, this is obvious; for arbitrary r, it is proved by induction by applying the operator Δ_{nt} on both sides of (6). From (6) we obtain:

$$\omega_r(f, nh) \leqslant n^r \omega_r(f, h). \tag{7}$$

As in Sec. 5, from this formula one derives (if $\lambda > 0$)

$$\omega_r(f, \lambda h) \leqslant (\lambda + 1)^r \omega_r(f, h). \tag{8}$$

There exist relations between the functions $\omega_r(f, h)$ of different orders. We shall restrict ourselves to a comparison of the moduli of continuity, $\omega(h)$ and $\omega_2(h)$, of a function $f \in C[A]$. Obviously,

$$\omega_2(h) \leqslant 2\omega(h). \tag{9}$$

In the opposite direction we have

THEOREM 9.   The following inequalities are true:

$$\omega(h) \leqslant \sum_{i=0}^{k} \frac{\omega_2(2^i h)}{2^{i+1}} + \frac{1}{2^{k+1}} \omega(2^{k+1}h), \qquad 2^k h \leqslant \tfrac{1}{3} l, \tag{10}$$

(Marchaud).[1] If $\omega_2(l) \neq 0$, then

$$\omega(h) \leqslant M\omega_2(\sqrt{h}), \qquad h \leqslant \tfrac{1}{4} l^2, \tag{11}$$

where the constant M depends upon f and l (Trigub).

Proof. Let x_1, x_2 be two points of A with $| x_2 - x_1 | \leqslant h$. We can complete this couple to a triple of the form x, $x + t$, $x + 2t$, where $t = \pm (x_1 - x_2)$ and x_1 and x_2 are x and $x + t$ (or $x + t$ and x). If $A = K$, there is no difficulty; if $A = [a, b]$, we must assume that $| t | \leqslant \tfrac{1}{3} l$. From

$$| f(x_2) - f(x_1) | \leqslant \tfrac{1}{2} | f(x + 2t) - 2f(x + t) + f(x) | + \tfrac{1}{2} | f(x) - f(x + 2t) |$$

we derive

$$\omega(h) \leqslant \tfrac{1}{2} \omega_2(h) + \tfrac{1}{2} \omega(2h), \qquad h \leqslant \tfrac{1}{3} l. \tag{12}$$

In the same way, $\omega(2h) \leqslant \tfrac{1}{2} \omega_2(2h) + \tfrac{1}{2} \omega(4h)$. We can substitute this bound into (12) for $\omega(2h)$. Repeating this process k times, we obtain (10). This inequality can be regarded as an estimation of $\omega(h)$ in terms of $\omega_2(h)$, since the last term in (10) is very small.

To obtain (11), assume first that $h \leqslant l_0 = \min (1, \tfrac{1}{3} l, l^2)$. We can select integers $k \geqslant 0$, $i_0 \geqslant 0$ in such a way that $2^k h \leqslant \tfrac{1}{3} l < 2^{k+1}h$ and $2^{i_0} \leqslant h^{-1/2} < 2^{i_0+1}$. Let $\sum_1$, $\sum_2$ be the portions of the sum (10) that correspond

---

[1] For another form of (10), see Problem 5.

to $i \leqslant i_0$ and $i > i_0$, respectively. For the first sum, we use the inequality $\omega_2(2^i h) \leqslant \omega_2(\sqrt{h})$; for the second, we use (7), according to which

$$\omega_2(2^i h) \leqslant (2^{i-i_0})^2\, \omega_2(\sqrt{h}).$$

This gives $\Sigma_1 \leqslant \omega_2(\sqrt{h})$ and

$$\Sigma_2 \leqslant 2^{-2i_0}\omega_2(\sqrt{h})\, 2^k \leqslant \tfrac{4}{3}\, l\omega_2(\sqrt{h}),$$

by the definition if i_0 and k. Finally, by (8),

$$\omega_2(l) \leqslant \left(\frac{l}{\sqrt{h}} + 1\right)^2 \omega_2(\sqrt{h}) \leqslant \frac{4l^2}{h}\, \omega_2(\sqrt{h}).$$

This gives the inequality $h \leqslant M_1\omega_2(\sqrt{h})$, where $M_1 = 4l^2/\omega_2(l)$. Because of the definition of k, the last term in (10) does not exceed

$$2^{-k-1}\omega(\tfrac{2}{3}\,l) \leqslant 3h\omega\, (\tfrac{2}{3}\,l)\, l^{-1} \leqslant M_2\omega_2(\sqrt{h}).$$

Substituting all this into (10), we obtain (11) for all values $0 \leqslant h \leqslant l_0$. By 5(4), we see that $\omega(h) \leqslant \mathrm{const}\, \omega(l_0) \leqslant \mathrm{const}\, \omega_2(\sqrt{l_0})$, for $h \geqslant l_0$. Hence, (11) holds in full generality. ∎

7. Classes of Functions

In this section we shall list several classes of functions, which will be constantly used later. A function f, defined on $A = [a, b]$ or $A = K$, satisfies a *Lipschitz condition* with constant M and exponent α, or belongs to the class $\mathrm{Lip}_M \alpha$, $M \geqslant 0$, $0 < \alpha \leqslant 1$, if

$$|f(x') - f(x)| \leqslant M\,|x' - x|^\alpha, \qquad x, x' \in A. \tag{1}$$

Equivalent to this is the inequality $\omega(f, h) \leqslant Mh^\alpha$. As an example, if f has a derivative that satisfies $|f'(x)| \leqslant M$, $x \in A$, then $f \in \mathrm{Lip}_M 1$. The class $\mathrm{Lip}\,\alpha$ consists of all functions that satisfy (1) for *some* M. It is easy to see that $\mathrm{Lip}\,\alpha \supset \mathrm{Lip}\,\beta$ if $0 < \alpha \leqslant \beta \leqslant 1$.

More general is the class $\Lambda_\omega(M)$, where $\omega(h)$ is a modulus of continuity: This consists of all functions $f \in C[A]$ with the property $\omega(f, h) \leqslant M\omega(h)$.

Other classes are based on continuity properties of the derivatives. Let $p = 0, 1, \cdots, M_i \geqslant 0$, $i = 0, \cdots, p + 1$, let ω be a modulus of continuity, and let $A = [a, b]$ or $A = K$. Then $\Lambda_{p\omega} = \Lambda_{p\omega}\,(M_0, \cdots, M_{p+1}; A)$ is the set of all functions $f \in C[A]$, which have continuous derivatives $f, f', \cdots, f^{(p)}$ on A, satisfying

$$|f^{(i)}(x)| \leqslant M_i, \qquad x \in A, \qquad i = 0, \cdots, p, \tag{2}$$

$$\omega(f^{(p)}, h) \leqslant M_{p+1}\omega(h). \tag{3}$$

We write $\Lambda_{p\alpha}$ for $\Lambda_{p\omega}$ if $\omega(h) = h^\alpha$.

If, in the preceding conditions, we allow the M_i to change with the function f, we obtain instead of $\Lambda_{p\omega}$ a linear space $X = X_{p\omega} : f \in X_{p\omega}$ means that $f, \cdots, f^{(p)}$ are continuous and that (3) is satisfied with *some* M_{p+1}. We can make $X_{p\omega}$ a *normed* linear space by defining $\| f \|_X$ to be the smallest possible $\max M_i$, $i = 0, \cdots, p + 1$, for the given f; in other words,

$$\| f \|_X = \max \left\{ \| f \|, \cdots, \| f^{(p)} \|, \max_h \frac{\omega(f^{(p)}, h)}{\omega(h)} \right\}. \tag{4}$$

Of course $\| f \|_X$ is different from the uniform norm $\| f \|$. One checks in a trivial fashion that (4) is a norm [compare 5(6)].

THEOREM 10.   (a) $\Lambda_{p\omega}$ is compact in $C[A]$; (b) $X_{p\omega}$ is a Banach space.

Proof.   (a)   Let $f_k \in \Lambda_{p\omega}$, $k = 1, 2, \cdots$, then the functions f_k are uniformly bounded, $|f_k(x)| \leqslant M_0$, and equicontinuous. If $p = 0$, this follows from (3) and if $p > 0$, from the uniform boundedness of the derivatives f_k'. By Arzelà's theorem [28, p. 128], there is a subsequence f_{k_i} that converges uniformly on A to a continuous function g. The same applies to the derivatives $f_k^{(j)}, j = 1, \cdots, p$. We can assume, replacing the sequence f_k by a subsequence if necessary, that $f_k^{(j)} \to g_j$ uniformly on A for $j = 1, \cdots, p$. By standard theorems of calculus, the g_j are the derivatives of g.

On the other hand, $\Lambda_{p\omega}$ is closed under the uniform convergence of derivatives of orders $j = 0, \cdots, p$, since the inequalities (2) and (3) remain preserved under this convergence. Hence, $g \in \Lambda_{p\omega}$.

(b)   If a sequence $f_k \in X_{p\omega}$, $k = 1, \cdots$, is a Cauchy sequence in this space (that is, if $\| f_k - f_l \|_X \to 0$ for $k, l \to \infty$), then for each $j = 0, \cdots, p$, $\| f_k^{(j)} - f_l^{(j)} \| \to 0$; hence, $\| f_k^{(j)} \| - \| f_l^{(j)} \| \to 0$ so that the sequence $\| f_k^{(j)} \|$, $k = 1, \cdots$, is bounded. The same applies to the sequence $\max_h [\omega(f_k^{(p)}, h)/\omega(h)]$. Hence, $f_k \in \Lambda_{p\omega}(M_0, \cdots, M_{p+1}; A)$, $k = 1, 2, \cdots$, for some constants M_j. By the proof of (a), $f_{k_i}^{(j)}(x) \to g^{(j)}(x)$ uniformly, $j = 0, \cdots, p$, for some $g \in X_{p\omega}$. Passing to the limit for $i \to \infty$ in the relations $|f_k^{(j)}(x) - f_{k_i}^{(j)}(x)| \leqslant \epsilon, j = 0, \cdots, p$, $x \in A$, $\omega(f_k^{(p)} - f_{k_i}^{(p)}, h) \leqslant \epsilon\omega(h)$, $0 \leqslant h \leqslant l$ (which are true for each $\epsilon > 0$ if i and k are sufficiently large), we obtain $\| f_k - g \|_X \leqslant \epsilon$. Thus, $X_{p,\omega}$ is complete. ▮

The class $W_p[A]$, $p = 1, 2, \cdots$, $A = [a, b]$ or $A = K$, consists of all functions $f \in C[A]$, which have an absolutely continuous derivative $f^{(p-1)}$ (so that $f^{(p)}(x)$ exists almost everywhere), with $|f^{(p)}(x)| \leqslant 1$ a.e. We write W_p^* for $W_p[K]$.

We also define classes $\Lambda_{p\omega}^s = \Lambda_{p\omega}^s(M_0, \cdots, M_{p+1}; A)$ of functions $f(x_1, \cdots, x_s)$ of s variables on an s-dimensional set A, which is either a parallelepiped or a torus (see Sec. 5). A function f belongs to $\Lambda_{p\omega}^s$ if and only if all its partial derivatives of orders $j = 0, \cdots, p$ exist and are continuous and satisfy the following conditions: For each partial derivative $D^j f$ of order j, $\| D^j f \| \leqslant M_j$, $j = 0, \cdots, p$, and in addition for each derivative of order p, $\omega(D^p f, h) \leqslant M_{p+1}\omega(h)$.

Theorem 10 holds also for the classes $\Lambda_{p\omega}^s$ and for the corresponding spaces $X_{p\omega}^s$. The class $\text{Lip}_M 1$ consists of all functions f on A that have the property $\omega(f, h) \leqslant Mh$ or

$$|f(y_1, \cdots, y_s) - f(x_1, \cdots, x_s)| \leqslant Mh \qquad \text{for} \qquad |y_k - x_k| \leqslant h, \quad k = 1, \cdots, s. \tag{5}$$

Returning to functions of one variable, we say that f is *quasi-smooth* if it satisfies $\omega_2(f, h) = O(h)$. *All functions of the class* Lip 1 *are quasi-smooth, but not conversely.* For example, the function

$$g(x) = \begin{cases} x \log |x|, & x \neq 0, \\ 0, & x = 0 \end{cases}; \qquad -1 \leqslant x \leqslant 1 \tag{6}$$

is quasi-smooth on $[-1, +1]$, but is not of class Lip 1. We first show that g is quasi-smooth on $[0, 1]$. Considering the second difference $\Delta_t^2 g(x)$, we may assume (by its symmetry about the midpoint $x + t$) that $t > 0$. Then

$$\Delta_t^2 g(x) = 2t \log \frac{x + 2t}{x + t} - x \log \left(1 + \frac{t}{x}\right) + x \log \left(1 + \frac{t}{x + t}\right).$$

Here, $1 \leqslant (x + 2t)/(x + t) \leqslant 2$, so that the first term on the right does not exceed $2t \log 2$. Using the inequality $\log(1 + u) \leqslant u$, $u \geqslant 0$, we see also that the two other terms are $O(t)$.

Next we observe that each function g *that is quasi-smooth on* $[0, 1]$ *and odd on* $[-1, +1]$, *is quasi-smooth on the larger interval.* Again we estimate the difference $\Delta_t^2 g(x)$, $t > 0$. The only interesting case is when not all three points x, $x + t$, and $x + 2t$ are in the same subinterval $[-1, 0]$ or $[0, 1]$. We may assume that the point x is in the first interval and that $x + t$, $x + 2t$ are in the second. Then $0 \leqslant -x \leqslant t$. We have

$$\Delta_t^2 g(x) = g(x + 2t) - 2g(x + t) - g(-x)$$

$$= [g(x + 2t) - 2g(t) + g(-x)] - 2[g(-x) - 2g(\tfrac{1}{2}t) + g(x + t)]$$

$$+ 2[g(t) - 2g(\tfrac{1}{2}t) + g(0)]$$

$$= O(x + t) + O\left(x + \frac{t}{2}\right) + O\left(\frac{t}{2}\right) = O(t).$$

On the circle K we have $f(x) = g(\pi x)$ as a nontrivial quasi-smooth function. Indeed, f is quasi-smooth on $[-\tfrac{1}{2}\pi, \tfrac{1}{2}\pi]$ and on $[\tfrac{1}{4}\pi, \tfrac{7}{4}\pi]$ (on the last interval because it is of class Lip 1 there); any three points x, $x + t$, $x + 2t$ of K with a sufficiently small t belong either to the first or the second interval.

8. Notes

1.   There are many generalizations and variations of Bernstein's inequalities 2(1), 2(4), 2(5). Let

$$T_n(x) = \sum_{k=1}^{n} (b_k \cos kx - a_k \sin kx)$$

be the conjugate of a trigonometric polynomial $T_n(x)$ with coefficients a_k, b_k. If $|\ T_n(x)\ | \leqslant M$, $x \in K$, then

$$T'_n(x)^2 + \widetilde{T}'_n(x)^2 \leqslant n^2 M^2; \qquad \widetilde{T}'_n(x)^2 + n^2 T_n(x)^2 \leqslant n^2 M^2.$$

The first of these inequalities is due to Szegö [1, p. 140].

2.   For special polynomials, one can sometimes improve the estimate $|\ P'_n(x)\ | \leqslant Mn^2$ of Markov's theorem. For example, Erdös [45] proves that if an algebraic polynomial $P_n(x)$ has only real roots that are not in the interval $[-1, +1]$, then, for this interval, $\|\ P'_n\ \| \leqslant \frac{1}{2} en\ \|\ P_n\ \|$. According to Lax [72], one can replace n by $\frac{1}{2} n$ in the inequality 2(6), if the polynomial P_n has no roots in $|\ z\ | < 1$.

Inequalities of the same type, but with a less precise constant, hold for the wider class of polynomials with positive coefficients in x, $1 - x$; that is, for the polynomials

$$P_n(x) = \sum_{k+l \leqslant n} a_{kl} x^k (1 - x)^l, \qquad a_{kl} \geqslant 0, \qquad 0 \leqslant x \leqslant 1.$$

(Lorentz [75]; see also Problem 7).

3.   An interesting parallel to interpolation by polynomials consists in interpolation by "spline functions," developed by Schoenberg, de Boor, and others; compare, for example, [88].

PROBLEMS

1.   Let $f(x) = e^x$, $-1 \leqslant x \leqslant 1$. Show that $E_n(f) \leqslant e2^{-n}/(n+1)!$; compare this with the approximation of e^x by the partial sums of its Taylor series.

2.   If $|\ P_n(z)\ | \leqslant M$ in $|\ z\ | \leqslant 1$, then $|\ P_n(z)\ | \leqslant Mr^n$ in $|\ z\ | \leqslant r$, where $r \geqslant 1$.

3.   The function $\omega(x) = x$ for $0 \leqslant x \leqslant \frac{1}{3} l$, $\omega(x) = \frac{1}{3} l$ for $\frac{1}{3} l \leqslant x \leqslant \frac{2}{3} l$, and $\omega(x) = x - \frac{1}{3} l$ for $\frac{2}{3} l \leqslant x \leqslant l$ is a modulus of continuity, although $\omega(x)/x$ is not monotone-decreasing.

4.   Prove that a function $f \in C[a, b]$ is linear if and only if $\omega_2(f, h) = 0$ for some $h > 0$.

5. Prove that

$$\omega(f, h) \leqslant 12h \left\{ \int_h^l \frac{\omega_2(f, t)}{t^2} \, dt + l^{-1} \| f \| \right\}$$

(Marchaud. *Hint:* Use 6(10)).

6. Show that 6(11) does not hold with a constant M independent of f, even if the functions f are assumed to be not linear.

7. If $P_n(z)$ is a polynomial with real coefficients, and has no roots in the circle $| z | \leqslant 1$, then either P_n or $-P_n$ is a polynomial with positive coefficients in x, $1 - x$ (abbreviated: ppc).

8. The integral $\int_0^x P_n(t) \, dt$ of a ppc P_n is again a ppc. Each polynomial is the derivative of a ppc.

9. Let P_{n-1} be the polynomial of best approximation of a k-times differentiable function f on $[-1, +1]$. Let $0 \leqslant k < n - 1$. Then, for some $n - k$ distinct points $x_1, \cdots, x_{n-k}$ of $(-1, +1)$, and some $\xi \in (-1, +1)$,

$$R_{nk}(x) = f^{(k)}(x) - P_{n-1}^{(k)}(x) = \frac{(x - x_1) \cdots (x - x_{n-k})}{(n - k)!} f^{(n)}(\xi).$$

(*Hint:* Use the remainder formula for the Lagrange interpolation, [4, p. 56]).

10. Let f be n-times continuously differentiable on $[-1, +1]$, $C_k = \| f^{(k)} \|$, $k = 0, \cdots, n$. Then

$$| R_{nk}(x) | \leqslant \frac{2^{n-k} C_n}{(n - 1)!}.$$

11. In the notations of Problem 10,

$$C_k \leqslant n^{2k} C_0 + \frac{2^{n-k}}{(n - k)!} C_n \qquad \text{(Gorny)}.$$

12. Prove that a function $f \in C^*$ belongs to Lip 1 if and only if $\| \sigma_n(f') \| = O(1)$, where $\sigma_n(f')$ are the arithmetic means of the (formally) differentiated Fourier series of f.

[4]

The Degree of Approximation by Trigonometric Polynomials

1. Generalities

The nth degree of approximation of a function $f \in C^*$ by trigonometric polynomials T_n of degree n is defined by

$$E_n^*(f) = \min_{T_n} \| f - T_n \|, \qquad \| f - T_n \| = \max_x | f(x) - T_n(x) | . \qquad (1)$$

The theorem of Weierstrass (Theorem 4, Chapter 1) tells us that $E_n^*(f) \to 0$ as $n \to \infty$ for each $f \in C^*$. Now we should like to estimate how fast $E_n^*(f)$ approaches zero. This information can be obtained if additional information about the function f is given; for example, if we know its modulus of continuity, the class Lip α to which it belongs, or the number of times it can be differentiated. In general, the "smoother" the function, the faster $E_n^*(f)$ tends to zero. Theorems that guarantee this are sometimes called *direct theorems* of approximation theory (Sec. 2 and 3). Conversely, *inverse theorems* (Secs. 4 and 5) assert that a function f has certain smoothness properties if $E_n^*(f)$ tends rapidly enough to zero. In exceptional but important cases, the direct and the inverse theorems match each other. In this way we are able to characterize the functions of the classes Lip α, $X_{p\alpha}$ by means of the order of magnitude of their degree of approximation.

In the search for a trigonometric polynomial T_n that is close to a given function f, it is natural to try some linear operators $L_n(f, x)$. The partial sums of the Fourier series $s_n(f, x)$, given in 1.2(9), do not converge uniformly for all f (Chapter 7, Sec. 2), and even their arithmetic means $\sigma_n(f, x)$, given in 1.2(12), do not always provide the best possible results. The loss of precision, however, is only slight. We have, for example,

THEOREM 1 (Lebesgue). There exists a constant M such that, for each $f \in C^*$,

$$\| f - s_n(f) \| \leqslant M E_n^*(f) \log n, \qquad n > 1. \qquad (2)$$

To prove this, recall that $s_n(f)$ is a linear operator of norm $\| s_n \| \leqslant A \log n$,

54

which leaves invariant each trigonometric polynomial of degree n (Chapter 1, Sec. 2). If T_n is the polynomial of best approximation to f, then

$$\| f - s_n(f) \| = \| (f - T_n) - s_n(f - T_n) \| \leqslant \| f - T_n \| + \| s_n(f - T_n) \|$$

$$\leqslant E_n^*(f)(1 + \| s_n \|) \leqslant M E_n^*(f) \log n.$$ ∎

2. The Theorem of Jackson

In this section we shall estimate the degree of approximation of a function by means of its moduli of continuity. We consider the following operator:

$$\int_{-\pi}^{\pi} f(t) L_n(x - t) \, dt = \int_{-\pi}^{\pi} f(x + t) L_n(t) \, dt, \tag{1}$$

where L_n is the *Jackson kernel*

$$L_n(t) = \lambda_n^{-1} \left(\frac{\sin (nt/2)}{\sin (t/2)} \right)^4, \qquad \int_{-\pi}^{\pi} L_n(t) \, dt = 1, \tag{2}$$

where the last relation defines λ_n .

We shall study some properties of L_n . Dividing the relation 1.2(11) by $\sin(\alpha/2)$ and using 1.2(10), we show that $(\sin (nt/2)/\sin (t/2))^2$ is a linear combination of 1, $\cos t$, $\cdots$, $\cos (n - 1) t$. Thus, $L_n(t)$ is the square of an even trigonometric polynomial of degree $n - 1$, that is, a positive, even trigonometric polynomial of degree $2(n - 1)$. We see that the integral (1) is also a trigonometric polynomial of degree $2(n - 1)$; it is even if the function f is even.

Next we prove that

$$\lambda_n \approx n^3. \tag{3}$$

Since $t/\pi \leqslant \sin (t/2) \leqslant \frac{1}{2} t$ for $0 \leqslant t \leqslant \pi$ (see p. 41), we have

$$\lambda_n \approx \int_0^{\pi} \left(\frac{\sin (nt/2)}{t/2} \right)^4 dt = 2n^3 \int_0^{(1/2)\pi n} \left(\frac{\sin u}{u} \right)^4 du \approx n^3.$$

Using (3) and the same method, we can prove that

$$\int_0^{\pi} t^k L_n(t) \, dt \approx n^{-k}, \qquad k = 0, 1, 2. \tag{4}$$

In fact, using (3) and the same inequality for $\sin (t/2)$, we infer that the last integral is $\approx n^{-k} \int_0^{(1/2)\pi n} u^{k-4} \sin^4 u \, du \approx n^{-k}$ for $0 \leqslant k \leqslant 2$.

It is convenient to normalize the operator (1) in such a way as to obtain a trigonometric polynomial of degree n. For this purpose, we put

$$K_n(t) = L_{n'}(t), \qquad n' = \left[\frac{n}{2} \right] + 1. \tag{5}$$

This is an even trigonometric polynomial of degree n, since $2(n' - 1) \leqslant n$. From (2) and (4),

$$\int_{-\pi}^{\pi} K_n(t)\,dt = 1, \qquad \int_0^{\pi} t^k K_n(t)\,dt \approx n^{-k}, \qquad k = 0, 1, 2. \tag{6}$$

The operator

$$J_n(x) = J_n(f, x) = \int_{-\pi}^{\pi} f(x + t)\,K_n(t)\,dt \tag{7}$$

will be called *the Jackson operator*.

THEOREM 2 (Jackson [58], [6]).   There exists a constant M such that, for each $f \in C^*$,

$$E_n^*(f) \leqslant M\omega\left(f, \frac{1}{n}\right), \qquad n = 1, 2, \cdots, \tag{8}$$

and even

$$E_n^*(f) \leqslant M\omega_2\left(f, \frac{1}{n}\right). \tag{9}$$

Proof.   Since, $\omega_2(f, h) \leqslant 2\omega(f, h)$ [see 3.6(9)], it is sufficient to prove (9). For this purpose, we estimate the difference $f(x) - J_n(x)$. Since K_n is even,

$$f(x) - J_n(x) = \int_{-\pi}^{\pi} [f(x) - f(x + t)]\,K_n(t)\,dt$$

$$= \int_0^{\pi} [2f(x) - f(x + t) - f(x - t)]\,K_n(t)\,dt.$$

The absolute value of the difference in the square bracket does not exceed

$$\omega_2(f, t) = \omega_2(t) = \omega_2\left(nt \cdot \frac{1}{n}\right) \leqslant (nt + 1)^2\,\omega_2\left(\frac{1}{n}\right)$$

[see 3.6(8)]. Therefore, for all x,

$$|f(x) - J_n(x)| \leqslant \omega_2\left(\frac{1}{n}\right) \int_0^{\pi} (nt + 1)^2\,K_n(t)\,dt.$$

From this and (6) we derive (9).   ∎

It should be noted that (8) or (9) cannot be obtained by using the operators $s_n(f)$ or $\sigma_n(f)$ instead of $J_n(f)$. The reason is that, for the Dirichlet and the Fejér kernels, there are no estimates similar to (6).

3. The Degree of Approximation of Differentiable Functions

We shall need a generalization of Jackson's kernel 2(2). This kernel and the corresponding constant $\lambda_n = \lambda_{nr}$ are defined by the relations

$$L_{nr}(t) = \lambda_{nr}^{-1} \left(\frac{\sin{(nt/2)}}{\sin{(t/2)}} \right)^{2r}, \qquad \int_{-\pi}^{\pi} L_{nr}(t)\, dt = 1; \tag{1}$$

L_{n1} is Fejér's kernel, $L_{n2} = L_n$ is Jackson's kernel.

We note some properties of L_{nr}. The kernel is a positive, even trigonometric polynomial of degree $r(n-1)$. Since $t/\pi \leqslant \sin{(t/2)} \leqslant \frac{1}{2}t$ for $0 \leqslant t \leqslant \pi$, we have, for each fixed r,

$$\lambda_n \approx \int_0^{\pi} \left(\frac{\sin{(nt/2)}}{t/2} \right)^{2r} dt = 2n^{2r-1} \int_0^{(1/2)\pi n} \left(\frac{\sin u}{u} \right)^{2r} du \approx n^{2r-1}. \tag{2}$$

We also have [compare the proof of 2(4)]

$$\int_0^{\pi} t^k L_{nr}(t)\, dt \approx n^{-k}, \qquad k = 0, 1, \cdots, 2r - 2. \tag{3}$$

The kernel

$$K_{nr}(t) = L_{n'r}(t), \qquad n' = \left[\frac{n}{r} \right] + 1 \tag{4}$$

is a positive, even trigonometric polynomial of degree n; from (1) and (3),

$$\int_{-\pi}^{\pi} K_{nr}(t)\, dt = 1; \qquad \int_0^{\pi} t^k K_{nr}(t)\, dt \approx n^{-k}, \qquad k = 0, \cdots, 2r - 2. \tag{5}$$

THEOREM 3. For each $p = 1, 2, \cdots$, there is a constant M_p with the property that if $f \in C^*$ has a continuous derivative $f^{(p)}$, then

$$E_n^*(f) \leqslant \frac{M_p}{n^p}\, \omega \left(f^{(p)}, \frac{1}{n} \right), \qquad n = 1, 2, \cdots. \tag{6}$$

Proof. We shall make use of the following operator:

$$I_n(x) = I_{np}(f, x) = -\int_{-\pi}^{\pi} K_{nr}(t) \sum_{k=1}^{p+1} (-1)^k \binom{p+1}{k} f(x + kt)\, dt, \tag{7}$$

where r is the smallest integer for which $r \geqslant \frac{1}{2}(p + 3)$. This integral is a linear combination of

$$\int_{-\pi}^{\pi} f(x + kt) \cos lt\, dt, \qquad l = 0, \cdots, n, \qquad k = 1, \cdots, p + 1. \tag{8}$$

We note that as a function of t, $f(x + kt)$ has period $2\pi/k$.

The following general remark will be useful. If $\Phi \in C^*$ has period $2\pi/k$, and if l is not divisible by k, then

$$\int_{-\pi}^{\pi} \Phi(t) \cos lt \, dt = \int_{-\pi}^{\pi} \Phi(t) \sin lt \, dt = 0. \tag{9}$$

In fact,

$$\int_0^{2\pi} \Phi(t) \, e^{ilt} dt = \int_{2\pi/k}^{2\pi+2\pi/k} \Phi(t) \, e^{ilt} dt = e^{2\pi il/k} \int_0^{2\pi} \Phi(t) \, e^{ilt} dt.$$

If l is divisible by k, the substitution $x + kt = u$ shows that (8) is a trigonometric polynomial in x of degree l/k. It follows that $I_n(x)$ is a trigonometric polynomial of degree n. Also, by 3.6(1), for each x:

$$| f(x) - I_n(x) | = \left| \int_{-\pi}^{\pi} K_{nr}(t) \, \Delta_t^{p+1} f(x) \, dt \right| \leqslant \int_{-\pi}^{\pi} K_{nr}(t) \omega_{p+1} \left(f, \mid t \mid \right) dt.$$

Since

$$\omega_{p+1}(h) \leqslant (nh + 1)^{p+1} \, \omega_{p+1} \left(\frac{1}{n} \right)$$

[see 3.6(8)], we obtain—using (5)—if $p + 1 \leqslant 2r - 2$, that is, if $r \geqslant \frac{1}{2}(p + 3)$,

$$| f(x) - I_n(x) | \leqslant 2\omega_{p+1} \left(\frac{1}{n} \right) \int_0^{\pi} (nt + 1)^{p+1} \, K_{nr}(t) \, dt$$

$$\leqslant M_p \omega_{p+1} \left(\frac{1}{n} \right), \qquad n = 1, 2, \cdots . \tag{10}$$

Thus, $E_n^*(f) \leqslant M_p \omega_{p+1}(1/n)$. Now, if $f^{(p)} \in C^*$, then (6) follows from 3.6(5). ∎

4. Inverse Theorems

In this and the next section we shall derive smoothness properties of a function $f \in C^*$ from the hypothesis that the numbers $E_n^*(f)$ approach zero with a given rapidity. Theorems of this kind were first obtained by Bernstein [35]. He also recognized the important role that his inequalities 3.2(1), and 3.2(5) play in connection with this question. His results have heen developed further by Zygmund [106], Stečkin [90], and others.

It is easy to see that if a function $\phi(n)$ is decreasing, then the sum $\sum\limits_{i=k}^{\infty} 2^i \phi(2^i)$ does not exceed $2 \sum\limits_{n > 2^{k-1}} \phi(n)$. Of the same type is the following more general inequality, which we shall use in this and the next chapter.

LEMMA 1. For each real p there is a constant M_p with the property that for each sequence $0 < u_k \leqslant u_{k+1} \leqslant \cdots \leqslant u_l$, such that $2 \leqslant u_i/u_{i-1} \leqslant 4$ for

$k < i \leqslant l$, and for each positive decreasing function $\phi(u)$ defined for $u \geqslant 0$ (or at least for all values u_i and all $u = 0, 1, \cdots$),

$$\sum_{i=k}^{l} u_i^p \phi(u_i) \leqslant M_p \sum_{[\frac{1}{2} u_k] \leqslant n < u_l} (n+1)^{p-1} \phi(n). \tag{1}$$

Proof. Assume first that $p \geqslant 1$. If $\phi(u)$ is not defined for all $u \geqslant 0$, we extend it by linear interpolation. We also define, for the purpose of the proof, $u_{k-1} = \frac{1}{2} u_k$. Then, since the function t^{p-1} is increasing,

$$\sum_{i=k}^{l} u_i^p \phi(u_i) \leqslant \sum_{i=k}^{l} (4u_{i-1})^{p-1} \phi(u_i)\, u_i$$

$$\leqslant 2 \cdot 4^{p-1} \sum_{i=k}^{l} u_{i-1}^{p-1} \phi(u_i)\, (u_i - u_{i-1})$$

$$\leqslant 2 \cdot 4^{p-1} \int_{u_{k-1}}^{u_l} t^{p-1} \phi(t)\, dt.$$

Since $t^{p-1} \phi(t) \leqslant (n+1)^{p-1} \phi(n)$ if $n \leqslant t \leqslant n+1$, the last integral does not exceed

$$\sum_{[u_{k-1}] \leqslant n < u_l} (n+1)^{p-1} \phi(n).$$

If $p < 1$, t^{p-1} decreases, and the proof is similar. In this case we assume, in addition, that $u_k \geqslant \mathrm{const} > 0$. ∎

THEOREM 4. There is a constant M such that, for each $f \in C^*$,

$$\omega(f, h) \leqslant Mh \sum_{0 \leqslant n \leqslant h^{-1}} E_n^*(f), \qquad h > 0. \tag{2}$$

There exist similar theorems concerning the moduli of continuity of higher orders. In fact, Theorem 4 is a special case, for $p = 1$, of the following

THEOREM 5. There exist constants M_p, $p = 1, 2, \cdots$, such that, for each function $f \in C^*$,

$$\omega_p(f, h) \leqslant M_p h^p \sum_{0 \leqslant n \leqslant h^{-1}} (n+1)^{p-1} E_n^*(f), \qquad h > 0. \tag{3}$$

Proof of Theorem 5. For each n, let T_n denote the trigonometric polynomial of best approximation to f, so that $\| f - T_n \| = E_n^*(f)$. If $h > 1$, (3) reduces to $\omega_p(f, h) \leqslant M_p h^p E_0^*(f)$; this is true because

$$\omega_p(f, h) = \omega_p(f - T_0, h) \leqslant 2^p \| f - T_0 \| = 2^p E_0^*(f),$$

by 3.6(3) and 3.6(1). We shall assume that $0 < h \leqslant 1$.

For the difference $\Delta_h^p f(x)$ we have, using 3.6(4), for $k = 0, 1, \cdots$,

$$| \Delta_h^p f(x) | \leqslant | \Delta_h^p T_{2^k}(x) | + 2^p E_{2^k}^*(f)$$

$$\leqslant h^p \| T_{2^k}^{(p)} \| + 2^p \phi(2^k), \tag{4}$$

where we have put $\phi(n) = E_n^*(f)$. We see that we have to estimate $\| T_{2^k}^{(p)} \|$. We have

$$T_{2^k}^{(p)}(x) = [T_1^{(p)}(x) - T_0^{(p)}(x)] + \sum_{i=1}^{k} [T_{2^i}^{(p)}(x) - T_{2^{i-1}}^{(p)}(x)]. \tag{5}$$

To estimate the terms of the last sum, we notice that

$$\| T_{2^i} - T_{2^{i-1}} \| \leqslant 2\phi(2^{i-1}), \qquad i = 1, 2, \cdots, \qquad \| T_1 - T_0 \| \leqslant 2\phi(0),$$

and apply Bernstein's inequality 3.2(5). This gives, according to Lemma 1,

$$\| T_{2^k}^{(p)} \| \leqslant 2\phi(0) + 2^{p+1} \sum_{i=1}^{k} (2^{i-1})^p \phi(2^{i-1})$$

$$\leqslant M_p' \sum_{0 \leqslant n < 2^{k-1}} (n + 1)^{p-1} \phi(n). \tag{6}$$

But

$$\sum_{1 \leqslant n \leqslant 2^k} (n + 1)^{p-1} \phi(n) \geqslant \phi(2^k) \sum_{1 \leqslant n \leqslant 2^k} (n + 1)^{p-1} \geqslant M_p'' 2^{kp}\phi(2^k), \qquad M_p'' > 0,$$

and from (4) we obtain

$$| \Delta_h^p f(x) | \leqslant \text{const} \left(h^p + 2^{-kp} \right) \sum_{n=0}^{2^k} (n + 1)^{p-1} \phi(n).$$

It is now clear that if we select $k = 0, 1, \cdots$, in such a way that $2^k \leqslant h^{-1} < 2^{k+1}$, we shall have (3). ∎

We shall consider some examples. Let

$$E_n^*(f) = O(n^{-\alpha}), \qquad 0 < \alpha \leqslant 1.$$

Theorem 4 gives

$$\omega(f, h) = O(1) h \sum_{1 \leqslant n \leqslant h^{-1}} n^{-\alpha} = O(1) h \int_0^{h^{-1}} x^{-\alpha}dx = O(h^\alpha), \qquad 0 < h \leqslant 1,$$

if $\alpha < 1$ and $\omega(f, h) = O(h \log (1/h))$ if $\alpha = 1$. Under the same assumptions, Theorem 5 gives $\omega_2(f, h) = O(h^\alpha)$, $0 < \alpha \leqslant 1$. Combining these facts with Theorem 2, we obtain, as important special cases,

THEOREM 6 (Bernstein [35]).   A function $f \in C^*$ belongs to the class Lip α, $0 < \alpha < 1$, if and only if

$$E_n^*(f) = O(n^{-\alpha}). \tag{7}$$

THEOREM 7 (Zygmund [106]).   A function $f \in C^*$ is quasi-smooth if and only if $E_n^*(f) = O(n^{-1})$.

Theorem 7 illustrates the fact that the moduli of continuity of higher orders rather than $\omega(f, h)$ are necessary to describe some approximation classes.

A few more special cases follow by simple computation:

(a) $E_n^*(f) = O(\log^{-k} n)$, $k > 0$ holds if and only if $\omega(f, h) = O(\log^{-k}(1/h))$.

(b)   If $E_n^*(f) = O(\psi(1/n))$, where $\psi(u)$, $0 < u \leqslant 1$, is a positive increasing function, then $\omega(f, h) = O(1) h \int_h^1 t^{-2} \psi(t) dt$ for small h.

5.   Differentiable Functions

If the $E_n^*(f)$ approach zero rapidly enough, we can prove the existence of the derivatives $f', f'', \cdots$, and we can also estimate their degrees of approximation.

THEOREM 8.   If, for some integer $p \geqslant 1$,

$$\sum_{n=1}^{\infty} n^{p-1} E_n^*(f) < +\infty, \tag{1}$$

then $f^{(p)}$ exists and is continuous, and its degree of approximation satisfies

$$E_n^*(f^{(p)}) \leqslant M_p \sum_{[n/2]}^{\infty} k^{p-1} E_k^*(f); \tag{2}$$

the constant M_p depends only on p.

Proof.   If $T_k(x)$ is the polynomial of best approximation of degree k for $f(x)$, then

$$f(x) - T_n(x) = \sum_{i=1}^{\infty} \{T_{2^i n}(x) - T_{2^{i-1} n}(x)\}. \tag{3}$$

This series converges uniformly. The series obtained by formal differentiation,

$$T_n^{(r)}(x) + \sum_{i=1}^{\infty} \{T_{2^i n}^{(r)}(x) - T_{2^{i-1} n}^{(r)}(x)\}, \qquad 1 \leqslant r \leqslant p, \tag{4}$$

also converge uniformly as a consequence of (1), since the norm of the ith term does not exceed [by 3.2(5)]

$$2^{ir}n^r \| T_{2^i n} - T_{2^{i-1}n} \| \leqslant 2 \cdot 2^{ir}n^r \phi(2^{i-1}n), \qquad \phi(n) = E_n^*(f), \qquad (5)$$

and the convergence of the series $\sum (2^{i-1}n)^r \phi(2^{i-1}n)$ follows from Lemma 1.

By known theorems about uniformly convergent series, $f^{(p)}(x)$ exists and is equal to the sum of the series (4). This proves the first part of the theorem.

For the second part, we can now write

$$\| f^{(p)} - T_n^{(p)} \| \leqslant \sum_{i=1}^{\infty} \| T_{2^i n}^{(p)} - T_{2^{i-1}n}^{(p)} \|$$

$$\leqslant \text{const} \sum_{i=1}^{\infty} (2^{i-1}n)^p \phi(2^{i-1}n) \leqslant \text{const} \sum_{[n/2]} k^{p-1} \phi(k),$$

by Lemma 1. This completes the proof.   ∎

Of course, once we know $E_n^*(f^{(p)})$ from (2), we can estimate the modulus of continuity of $f^{(p)}$ by means of Theorem 4. We shall restrict ourselves to the simplest special case (see Theorem 3).

THEOREM 9.   A 2π-periodic continuous function f belongs to $X_{p\alpha}$ $p = 0, 1, \cdots$, $0 < \alpha < 1$, if and only if

$$E_n^*(f) = O\left(\frac{1}{n^{p+\alpha}}\right). \qquad (6)$$

6. Notes

1.   Let $\psi(u) > 0$ be an increasing function defined for $0 < u \leqslant 1$. In generalizing Theorem 6, one could ask, "For which ψ is it true that $E_n^*(f) = O(\psi(n^{-1}))$ is equivalent to $\omega(f, h) = O(\psi(h))$?" The answer to this has been given by Lozinskiĭ and by Bari and Stečkin [33]. It consists in the existence of a constant $C > 1$ for which

$$1 < \varliminf_{u \to 0} \frac{\psi(Cu)}{\psi(u)} \leqslant \varlimsup_{u \to 0} \frac{\psi(Cu)}{\psi(u)} < C.$$

2.   Let B be the space of all real (or complex) functions that are uniformly continuous and bounded on the real axis R: $-\infty < x < +\infty$, with the uniform norm. Neither the algebraic polynomials, unbounded on R, nor the 2π-periodic trigonometric polynomials approximate all functions of B. In 1946, Bernstein ([2, vol. II, pp. 371-394]) indicated the appropriate tool of approximation.

An entire function $f(z) = \sum\limits_{k=0}^{\infty} c_k z^k$ of the complex variable z is of *degree* $\sigma \geq 0$ (in the terminology of Bernstein) if it satisfies one of the equivalent conditions:

(a) $\overline{\lim\limits_{k\to\infty}} \sqrt[k]{k! \, |\, c_k\,|} \leq \sigma$;

(b) $\overline{\lim\limits_{r\to\infty}} \dfrac{\log M(r)}{r} \leq \sigma$, where $M(r) = \max\limits_{|z|=r} |\, f(z)\,|$;

(c) $|\, f(z)\,| \leq e^{(\sigma+\epsilon)|z|}$, for each $\epsilon > 0$ and all sufficiently large $|\, z\,|$.

If B_σ is the class of all entire functions of degree σ that are bounded on R, we define $A_\sigma(f)$—the degree of approximation of $f \in B$ by $g_\sigma \in B_\sigma$—as the minimum of $\|f - g_\sigma\|$ for all $g_\sigma \in B_\sigma$. For $A_\sigma(f)$, one has theorems similar to those developed in this chapter for trigonometric approximation. We quote as simplest examples: $A_\sigma(f) \leq \text{const } \omega(f, 1/\sigma)$; if $A_\sigma(f) \leq \sigma^{-\alpha}$, $0 < \alpha < 1$, then $f \in \text{Lip } \alpha$. It should be noted that a trigonometric polynomial of degree n belongs to B_n; conversely, if $g \in B_\sigma$ is 2π-periodic, then g is a trigonometric polynomial of degree $[\sigma]$ (see [1], [15]).

PROBLEMS

1. If $(a_0/2) + \sum\limits_{k=1}^{\infty} a_k \cos kx$ is the Fourier series of the function $f \in C^*$, then

$$\tfrac{1}{2}|\, a_{n+1}\,| \leq E_n^*(f) \leq \sum\limits_{k=n+1}^{\infty} |\, a_k\,|.$$

2. The Fourier series of the function $f \in C^*$ is uniformly convergent if $\omega(f, h) \log(1/h) \to 0$ for $h \to 0$ (Dini).

3. Let $\bar{E}_n(f)$ denote the degree of approximation of a function $f \in C^*$ by trigonometric polynomials without the constant term. There is a constant M such that $\bar{E}_n(f) \leq M\omega(f, n^{-1})$ for all f with mean-value zero.

4. If a function $f \in C^*$ has a continuous derivative f', then $E_n(f) \leq Mn^{-1}\bar{E}_n(f')$. [*Hint:* Consider the modulus of continuity of $f - T_n$, where T_n' approximates f'.]

5. Derive Theorem 3 from Theorem 2.

6. Using the method of proof of Theorem 2, show that $\|f - \sigma_n(f)\| \leq \text{const } n^{-\alpha}$ for $f \in \text{Lip } \alpha$, $0 < \alpha < 1$.

7. If $f \in C^*$ has a continuous derivative, then $nE^*(f) \to 0$. If $\sum\limits_{n=1}^{\infty} E_n^*(f) < +\infty$, then f has a continuous derivative.

8. Estimating $N^{-1}\sum\limits_{0}^{N} E_n^*(f)$ from below and $N^{-1}M^{-1}\sum\limits_{0}^{NM} E_n^*(f)$ from above, prove that if $\omega(f, h) \sim h^\alpha$ for $h \to 0$, then $E_n^*(f) \sim n^{-\alpha}$ $(0 < \alpha < 1)$.

9. Prove that if $\omega(f', h) \sim h^\alpha$, $0 < \alpha < 1$, then $E_n^*(f) \sim n^{-1-\alpha}$. Also prove the statement of Problem 8 for $\alpha = 1$.   .

10. Assume that Theorem 2 is known to be true if $f \in \text{Lip } 1$. Derive from this the statement (8) of Theorem 2 by using approximation of f by piecewise linear functions, but without the use of integral operators.

11. If $f \in C^*$ has a continuous second derivative, then for the Jackson polynomial J_n ,

$$\lim_{n \to \infty} n^2 [J_n(f, x) - f(x)] = \tfrac{3}{2} f''(x) \qquad \text{(Natanson)}.$$

[5]

The Degree of Approximation
by Algebraic Polynomials

1. Preliminaries

Let f be a continuous function on $[-1, +1]$ with the modulus of continuity $\omega(h)$. It is easy enough to find a rough estimate of the order of approximation of f by polynomials P_n of degree n. We consider the function $g(t) = f(\cos t)$, which belongs to C^*. If t changes by an amount h, $\cos t$ changes by at most h. Consequently, the modulus of continuity of g does not exceed $\omega(h)$. By Theorem 2, Chapter 4, there are trigonometric polynomials $T_n(t)$ with

$$|g(t) - T_n(t)| \leqslant M\omega\left(\frac{1}{n}\right);$$

from the proof of this theorem, one sees that for an even function $g(t)$, the $T_n(t)$ can also be taken even. But then $T_n(t)$ is nothing but an algebraic polynomial $P_n(x)$ of the variable $x = \cos t$. Thus, we obtain

$$|f(x) - P_n(x)| \leqslant M\omega\left(\frac{1}{n}\right), \qquad -1 \leqslant x \leqslant 1. \tag{1}$$

But if we apply the same substitution $x = \cos t$ to the inverse theorems of Chapter 4 (see Theorem 6), we obtain only that $E_n(f) = O(n^{-\alpha})$, $0 < \alpha < 1$ implies $f \in \mathrm{Lip}_M \alpha$ for each interval $[-1 + \delta, 1 - \delta]$, $\delta > 0$, with an M that increases to $+\infty$ as $\delta \to 0$. Consequently, proceeding in this way, we cannot characterize completely the classes $\mathrm{Lip}\,\alpha$ on $[-1, +1]$. The reason for this phenomenon has been discovered by Nikol'skiĭ [84]. He showed that the quality of approximation by polynomials *increases toward the end points of the interval*. Later, Timan [96] obtained the following results:

THEOREM 1.   There exists a constant M such that for each function $f \in C[-1, +1]$ there is a sequence of polynomials $P_n(x)$ for which

$$|f(x) - P_n(x)| \leqslant M\omega(f, \Delta_n(x)), \qquad -1 \leqslant x \leqslant 1, \qquad n = 0, 1, \cdots, \tag{2}$$

where $\Delta_n(x)$ are defined by

$$\Delta_n = \Delta_n(x) = \max\left(\frac{\sqrt{1 - x^2}}{n}, \frac{1}{n^2}\right), \qquad n = 1, 2, \cdots, \tag{3}$$

$$\Delta_0(x) = 1.$$

65

Thus, (2) is an essential improvement over (1) if x is close to -1 or $+1$. The functions $\Delta_n(x)$ or the similar functions $\sqrt{1 - x^2}/n$ often appear in the theory of approximation by algebraic polynomials. For further examples, see Theorem 3 in this Chapter, and Theorem 11 in Chapter 7.

THEOREM 2.   If the function f has continuous derivatives $f', \cdots, f^{(p)}$ on $[-1, +1]$, then there is a sequence of polynomials $P_n(x)$ for which

$$
|f(x) - P_n(x)| \leqslant M_p \Delta_n(x)^p \, \omega(f^{(p)}, \Delta_n(x)),
$$
$$
-1 \leqslant x \leqslant 1, \qquad n = p, p+1, \cdots;
$$

$$(4)$$

the constant M_p depends only upon p.

These theorems cannot be improved: Dzjadyk [43] obtained their converses for the cases when $f \in \text{Lip } \alpha$ or $f^{(p)} \in \text{Lip } \alpha$ (see Sec. 4).

The proofs of Theorems 1 and 2 will be based on the substitution $x = \cos t$, but in a more subtle way than has been attempted above. We shall need the following *generalization of Jackson's operator* [see 4.2(7)]:

$$
J_n(t) = J_{nr}(t) = \int_{-\pi}^{\pi} f(t + u) \, K_{nr}(u) \, du,
$$

$$(5)$$

where the kernel $K_n = K_{nr}$ is the same as in Chapter 4, Sec. 3:

$$
K_{nr}(t) = L_{n'r}(t), \qquad n' = \left[\frac{n}{r}\right] + 1,
$$

$$
L_{nr}(t) = \lambda_n^{-1} \left(\frac{\sin(nt/2)}{\sin(t/2)} \right)^{2r},
$$

$$(6)$$

$$
\int_{-\pi}^{\pi} L_{nr}(t) \, dt = 1; \qquad n, r = 1, 2, \cdots.
$$

In particular for $r = 1$, we obtain the operator $\sigma_{n+1}(f)$ given in 1.2(12), and for $r = 2$, Jackson's operator given in 4.2(7). We know that K_{nr} and therefore J_{nr} are trigonometric polynomials of degree n; J_{nr} is even if the function f is even.

The main lemma about K_{nr} concerns positive functions $\phi(u)$, $u \geqslant 0$, which increase with u and satisfy, for some positive integer m, the condition

$$
\phi(\lambda u) \leqslant (\lambda + 1)^m \phi(u), \qquad \lambda \geqslant 0, \qquad u \geqslant 0.
$$

$$(7)$$

For example, functions of the type $\phi(u) = u^p \omega(u)$, where ω is a modulus of continuity, satisfy this with $m = p + 1$ [see 3.5(4)].

The following technical lemma embodies almost all the calculations needed in Secs. 2 and 3.

LEMMA 1. There exist constants B_r, $r = 1, 2, \cdots$, with the following property. If ϕ and m have the formulated properties and s is a positive integer for which $2s + m \leqslant 2r - 2$, then for $n = 1, 2, \cdots$, $-\pi \leqslant t \leqslant \pi$,

$$\int_{-\pi}^{\pi} |\cos t - \cos (t + u)|^s \phi \left(\delta_n(t) + \frac{|u|}{n} \right) K_{nr}(u) \, du \leqslant B_r \delta_n(t)^s \phi(\delta_n(t)), \qquad (8)$$

where $\delta_n(t)$ is a function of t that is obtained from $\Delta_n(x)$ by the substitution $x = \cos t$:

$$\delta_0 = 1, \qquad \delta_n = \delta_n(t) = \max \left(\frac{|\sin t|}{n}, \frac{1}{n^2} \right), \qquad n = 1, 2, \cdots . \qquad (9)$$

Proof. We have

$$|\cos t - \cos (t + u)| = 2 \left| \sin \frac{u}{2} \sin \left(t + \frac{u}{2} \right) \right| \leqslant 2 \sin^2 \frac{u}{2} + 2 \left| \sin \frac{u}{2} \sin t \right|$$

$$\leqslant u^2 + |u| \, |\sin t| . \qquad (10)$$

Also, from (7) with

$$\lambda = \frac{\delta_n + \dfrac{|u|}{n}}{\delta_n} ,$$

we have

$$\phi \left(\delta_n + \frac{|u|}{n} \right) \leqslant \left(2 + \frac{|u|}{n\delta_n} \right)^m \phi(\delta_n). \qquad (11)$$

This gives, as an upper bound for the integral (8):

$$2\delta_n^s \phi(\delta_n) \int_0^{\pi} \left(\frac{u^2 + u \, |\sin t|}{\delta_n} \right)^s \left(2 + \frac{u}{n\delta_n} \right)^m K_n(u) \, du. \qquad (12)$$

Expanding both the terms in parentheses by the binomial formula, we see that the last integral does not exceed a finite sum of constant multiples of terms of the type

$$I = \int_0^{\pi} \left(\frac{u^2}{\delta_n} \right)^i \left(\frac{u \, |\sin t|}{\delta_n} \right)^j \left(\frac{u}{n\delta_n} \right)^k K_n(u) \, du, \qquad (13)$$

where $i + j = s$, $k \leqslant m$. Since $2i + j + k \leqslant 2r - 2$, we can apply 4.3(5) and obtain

$$I \approx \left(\frac{1}{n^2 \delta_n} \right)^{i+k} \left(\frac{|\sin t|}{n\delta_n} \right)^j \leqslant 1.$$

It follows that the integral (12) does not exceed a constant $B(s, m)$. If r is given, then s and m assume only a finite number of values, so that we can take B_r equal to the maximum of the constants $B(s, m)$ for all s, m. $\quad\blacksquare$

2. The Approximation Theorems

We can now prove our theorems.

Proof of Theorem 1.   Let $f \in C[-1, +1]$.   We show that

$$| f(\cos t) - J_n(t) | \leqslant M\omega(\delta_n(t)), \tag{1}$$

where M is a constant and J_n is the Jackson polynomial of the function $f(\cos t)$, given by 4.2(7) or 1(5) with $r = 2$. We have

$$f(\cos t) - J_n(t) = \int_{-\pi}^{\pi} [f(\cos t) - f(\cos (t + u))] K_n(u) \, du. \tag{2}$$

The absolute value of the difference in the square brackets does not exceed

$$\omega(| \cos t - \cos (t + u) |) \leqslant \left[\frac{| \cos t - \cos (t + u) |}{\delta_n} + 1 \right] \omega(\delta_n),$$

by 3.5(4). Therefore,

$$| f(\cos t) - J_n(t) | \leqslant \omega(\delta_n) \int_{-\pi}^{\pi} \left[\frac{| \cos t - \cos (t + u) |}{\delta_n} + 1 \right] K_n(u) \, du,$$

and (1) follows from Lemma 1 with $r = 2$, $s = 1$, $\phi(u) \equiv 1$, $m = 0$.

Since J_n is an even trigonometric polynomial of degree n, (1) gives $| f(x) - P_n(x) | \leqslant M\omega(\Delta_n)$, $n = 1, 2, \cdots$. We also have

$$| f(x) - f(0) | \leqslant \omega(1) = \omega(\Delta_0). \qquad \blacksquare$$

LEMMA 2.   Let B_r, $r = 1, 2, \cdots$, be the constants of Lemma 1. Let ϕ be a positive increasing function that satisfies 1(7), and let $2r \geqslant m + 4$. If, for some $n = 1, 2, \cdots$, the function f has a continuous derivative with

$$| f'(\cos t) | \leqslant \phi(\delta_n(t)) \tag{3}$$

for all real t, then for the polynomial J_{nr} of 1(5), with $f(t)$ replaced by $f(\cos t)$,

$$| f(\cos t) - J_n(t) | \leqslant B_r \delta_n(t) \, \phi(\delta_n(t)). \tag{4}$$

Proof.   We write the difference (4) in form (2). This time we have

$$| f(\cos t) - f(\cos (t + u)) | = | \cos t - \cos (t + u) | \, | f'(\cos \xi) |, \tag{5}$$

where $\xi = t + \theta u, 0 \leqslant \theta \leqslant 1$. We majorize $f'(\cos \xi)$ by $\phi(\delta_n(\xi))$. Now

$$\delta_n(\xi) = \max \left\{ \frac{| \sin \xi |}{n}, \frac{1}{n^2} \right\} \leqslant \max \left\{ \frac{| \sin t |}{n} + \frac{| \sin \theta u |}{n}, \frac{1}{n^2} \right\}$$

$$\leqslant \delta_n(t) + \frac{| u |}{n}. \tag{6}$$

Hence,

$$|f(\cos t) - J_n(t)| \leqslant \int_{-\pi}^{\pi} |\cos t - \cos(t+u)| \, \phi\left(\delta_n + \frac{|u|}{n}\right) K_n(u) \, du.$$

We can apply Lemma 1 with $s = 1$ to obtain (4).   ∎

LEMMA 3.   There exist constants B'_m, $m = 0, 1, \cdots$, with the following property: Let f have a continuous derivative on $[-1, +1]$, and for a positive increasing function ϕ of type 1 (7), let

$$|f'(x) - P_n(x)| \leqslant \phi(\Delta_n(x)), \qquad -1 \leqslant x \leqslant 1, \tag{7}$$

where P_n is some polynomial of degree $n = 0, 1, \cdots$. Then there is a polynomial Q_{n+1} of degree $n + 1$ with

$$|f(x) - Q_{n+1}(x)| \leqslant B'_m \Delta_{n+1}(x) \, \phi(\Delta_{n+1}(x)), \qquad -1 \leqslant x \leqslant 1. \tag{8}$$

Proof.   First assume that $n \geqslant 1$. Let P_{n+1} be an integral of P_n and let r be the smallest integer for which $2r \geqslant m + 4$. The function $g = f - P_{n+1}$ satisfies (3), and we may deduce (4). Then $|f(x) - Q_{n+1}(x)| \leqslant B_r \Delta_n \phi(\Delta_n)$, where $Q_{n+1}(x) = J_n(t) + P_{n+1}(x)$, $\cos t = x$, is a polynomial of degree $n + 1$ in x. It remains to observe that $\Delta_n \leqslant 4\Delta_{n+1}$. For $n = 0$, (8) follows from (7) by integration.   ∎

Theorem 2 follows from Theorem 1 by several applications of Lemma 3.   ∎

3.   Inequalities for the Derivatives of Polynomials

For the proofs of the inverse theorems of the next section, we need inequalities (due to Dzjadyk [43] and Brudnyĭ) that are more general than the inequality of Bernstein, 3.2(4). As in Sec. 2, the computations will be first carried out for trigonometric polynomials and the Jackson kernels, and later will be translated into statements concerning algebraic polynomials. We begin by finding an integral representation for the derivative of a trigonometric polynomial T_n of degree n by means of the polynomial itself.

A simple formula of this type follows from 1.2(9). Since the operator $s_n(f)$ preserves trigonometric polynomials of degree n, we have

$$T_n(t) = \pi^{-1} \int_{-\pi}^{\pi} T_n(u) \, D_n(u - t) \, du, \qquad D_n(u) = \tfrac{1}{2} + \cos u + \cdots + \cos nu.$$

Differentiating, we obtain

$$T'_n(t) = -\pi^{-1} \int_{-\pi}^{\pi} T_n(t + u) \, D'_n(u) \, du,$$

$$D'_n(u) = -(\sin u + \cdots + n \sin nu). \tag{1}$$

The absolute value of the trigonometric polynomial $\pi^{-1} D'_n(u)$ does not exceed n^2.

More generally, we shall prove for $r = 0, 1, \cdots, n = 1, 2, \cdots$:

$$T'_n(t) = 2^{2r} n^{-2r+2} \int_{-\pi}^{\pi} T_n(t + u)\, H_{nr}(u)\, \tilde{K}_{nr}(u)\, du,$$

$$\tilde{K}_{nr}(u) = \left(\frac{\sin{(nu/2)}}{\sin{(u/2)}}\right)^{2r},$$

$$(2)$$

where $H_{nr}(u)$ is some fixed trigonometric polynomial with the property $|H_{nr}(u)| \leqslant 1$; the kernel $\tilde{K}_{nr}$ is related in an obvious way to L_{nr}; see 1(6).

For $r = 0$, (2) follows from (1), so that we can prove (2) by induction in r. We assume that for some r and all $n = 1, 2, \cdots$, the polynomials H_{nr} exist. If $S_{2n}(t)$ is some trigonometric polynomial of degree $2n$, then by (2),

$$S'_{2n}(0) = 2^{2r} (2n)^{-2r+2} \int_{-\pi}^{\pi} S_{2n}(u)\, H_{2n,r}(u)\, \tilde{K}_{2n,r}(u)\, du. \qquad (3)$$

In particular, we can take

$$S_{2n}(t) = T_n(x + t) \left(\frac{\sin{(nt/2)}}{\sin{(t/2)}}\right)^2,$$

where T_n is an arbitrary polynomial of degree n, because the second factor is a polynomial of degree $n - 1$. Since

$$S'_{2n}(0) = n^2 T'_n(x) \qquad \text{and} \qquad \tilde{K}_{2n,r}(u) = 2^{2r} (\cos \tfrac{1}{2} nu)^{2r}\, \tilde{K}_{nr}(u),$$

relation (3) becomes

$$T'_n(x) = 2^{2(r+1)} n^{-2r} \int_{-\pi}^{\pi} T_n(x + u)\, H_{n,r+1}(u)\, \tilde{K}_{n,r+1}(u)\, du,$$

where

$$H_{n,r+1}(u) = H_{2n,r}(u)\, (\cos \tfrac{1}{2} nu)^{2r}$$

is also a trigonometric polynomial whose absolute value does not exceed 1.

We shall need a slight generalization of the function $\Delta_n(x)$ of 1(3). For $0 < a \leqslant 1, |x| \leqslant a$, we put

$$\Delta_n(x, a) = \max\left(\frac{\sqrt{a^2 - x^2}}{n}, \frac{1}{n^2}\right), \qquad n = 1, 2, \cdots, \qquad \Delta_0(x, a) = 1.$$

By means of the representation (2) we can prove

LEMMA 4.   Let $0 < a \leqslant 1, 0 < q < a$, and let ϕ be a positive increasing function that satisfies 1(7). There exists a constant $M = M(a, q, m)$ with the following property: If, for an algebraic polynomial P_n of degree n,

$$|P_n(x)| \leqslant \phi(\Delta_n(x, a)), \qquad |x| \leqslant a, \qquad (4)$$

then, with $a_1 = a - qn^{-2}$,

$$| P_n'(x) | \leqslant M \Delta_n(x, a_1)^{-1} \phi(\Delta_n(x, a_1)), \qquad | x | \leqslant a_1. \tag{5}$$

Proof. We put $x = a \cos t$. Since $\Delta_n(x, a) \leqslant \delta_n(t)$, we see that $T_n(t) = P_n(a \cos t)$ is a trigonometric polynomial of degree n, which satisfies $| T_n(t) | \leqslant \phi(\delta_n(t))$. We select r so that $m \leqslant 2r - 2$ and estimate T_n' by means of (2); we obtain

$$| P_n'(x) | \, | a | \sin t | \leqslant 2^{2r} n^{-2r+2} \int_{-\pi}^{\pi} \phi(\delta_n(t + u)) \, \tilde{K}_{nr}(u) \, du.$$

But, by 1(6), $\tilde{K}_{nr} = \lambda_n L_{nr} = \lambda_n K_{Nr}$, $N = r(n - 1)$, and $\delta_n(t) \approx \delta_N(t)$, uniformly. Therefore,

$$| P_n'(x) | \, | a | \sin t | \leqslant \text{const } n \int_{-\pi}^{\pi} \phi\left(\delta_n(t) + \frac{| u |}{n}\right) K_{Nr}(u) \, du$$

$$\leqslant \text{const } n\phi(\delta_N(t)) \leqslant \text{const } n\phi(\delta_n(t)). \tag{6}$$

We have used 4.3(2) here, the inequality $\delta_n(t + u) \leqslant \delta_n(t) + | u |/n$, and Lemma 1 with $s = 0$. From the definition of a_1, it follows that

$$\sqrt{a^2 - a_1^2} \approx \frac{1}{n}. \tag{7}$$

Hence, for $| x | \leqslant a_1$,

$$a | \sin t | = \sqrt{a^2 - x^2} \leqslant \sqrt{a^2 - a_1^2} + \sqrt{a_1^2 - x^2} \leqslant \sqrt{a_1^2 - x^2} + \text{const } n^{-1},$$

and therefore $\delta_n(t) \leqslant \text{const } \Delta_n(x, a_1)$. Hence, from 1(7),

$$\phi(\delta_n(t)) \leqslant \text{const } \phi(\Delta_n(x, a_1)), \qquad | x | \leqslant a_1.$$

Also, $a | \sin t | \geqslant \sqrt{a^2 - a_1^2} \geqslant \text{const } n^{-1}$, so that $n | \sin t |^{-1} \leqslant \text{const } n^2$. On the other hand, $a | \sin t | \geqslant \sqrt{a_1^2 - x^2}$. This gives

$$\frac{n}{| \sin t |} \leqslant \text{const min} \left\{ n^2, \frac{n}{\sqrt{a_1^2 - x^2}} \right\} = \frac{\text{const}}{\Delta_n(x, a_1)}, \qquad | x | \leqslant a_1.$$

Therefore, (6) implies (5). ∎

THEOREM 3. Let $r = 0, 1, \cdots$, and let ω be a modulus of continuity. If an algebraic polynomial P_n of degree n satisfies

$$| P_n(x) | \leqslant \Delta_n(x)^r \, \omega(\Delta_n(x)), \qquad | x | \leqslant 1, \tag{8}$$

then with a constant M_r that depends only on r,

$$| P_n'(x) | \leqslant M_r \Delta_n(x)^{r-1} \, \omega(\Delta_n(x)), \qquad | x | \leqslant 1. \tag{9}$$

Proof. We can assume that $n > 1$, for otherwise (9) follows trivially (or from Markov's theorem), since $\Delta_0 = \Delta_1 = 1$. The function $\phi(u) = u^r \omega(u)$ satisfies 1(7), with $m = r + 1$. If we take $a = 1$ in (6), we obtain

$$| P_n'(x) | \leqslant M' \frac{n}{\sqrt{1 - x^2}} \Delta_n(x)^r \, \omega(\Delta_n(x)), \qquad | x | \leqslant 1,$$

and this is identical with (9) if $\sqrt{1 - x^2} \geqslant n^{-1}$. It remains to establish (9) when $\sqrt{1 - x^2} < n^{-1}$. By Lemma 4,

$$| P_n'(x) | \leqslant M \Delta_n(x, a_1)^{r-1} \, \omega(\Delta_n(x, a_1)), \qquad | x | \leqslant a_1 = 1 - qn^{-2}. \tag{10}$$

We select $q = (r + 1)^{-1}$. If $r \geqslant 1$, we can apply Lemma 4 once more. After $j = 1, 2, \cdots, r + 1$ applications of Lemma 4, we shall have

$$| P_n^{(j)}(x) | \leqslant M' \Delta_n(x, a_j)^{r-j} \, \omega(\Delta_n(x, a_j)), \qquad | x | \leqslant a_j, \qquad j = 1, \cdots, r + 1, \tag{11}$$

where $a_j = 1 - jqn^{-2}$. We put $b = a_{r+1} = 1 - n^{-2}$. It is not quite true that the function $\omega(u)/u$ is decreasing. We have, however, from 3.5(4), $\omega(u')/u' \leqslant 2\omega(u)/u$ for $0 < u \leqslant u'$. The smallest value of $\Delta_n(x, b)$ for $| x | \leqslant b$ is equal to n^{-2} for $x = b$. Hence, from (11), with $j = r + 1$,

$$| P_n^{(r+1)}(x) | \leqslant 2M' n^2 \omega(n^{-2}), \qquad | x | \leqslant 1 - n^{-2}. \tag{12}$$

This upper bound of $P_n^{(r+1)}$ is a constant. From 3.4.2,

$$| P_n^{(r+1)}(x) | \leqslant M'' n^2 \omega(n^{-2}), \qquad | x | \leqslant 1. \tag{13}$$

Now we observe that (11), with $j = r$, assures the existence of a point in $(1 - n^{-2}, 1)$—namely, $x = a_r$—where the absolute value of $P_n^{(r)}(x)$ does not exceed $M' \omega(n^{-2})$. Since

$$P_n^{(r)}(x) = P_n^{(r)}(a_r) + \int_{a_r}^{x} P_n^{(r+1)}(t) \, dt,$$

it follows from (13) that $| P_n^{(r)}(x) | \leqslant \text{const } \omega(n^{-2})$ on $[1 - n^{-2}, 1]$, and similarly on $[- 1, - 1 + n^{-2}]$. After r integrations, we obtain

$$| P_n'(x) | \leqslant M_r n^{-2(r-1)} \omega(n^{-2}), \qquad 1 - n^{-2} \leqslant | x | \leqslant 1. \tag{14}$$

But if $\sqrt{1 - x^2} < n^{-1}$, then $1 - n^{-2} \leqslant | x |$, and also $\Delta_n(x) = n^{-2}$. This proves (9) for the remaining values of x.   ∎

4. Inverse Theorems

The next theorem corresponds to Theorem 4 of Chapter 4. The proof is similar but somewhat more delicate because of the presence of the function $\Delta_n(x)$.

THEOREM 4.   There is a constant M with the following property: If ω is a modulus of continuity, and if for a function $f \in C[-1, +1]$ and polynomials P_n,

$$|f(x) - P_n(x)| \leqslant \omega(\Delta_n(x)); \qquad n = 0, 1, \cdots, \qquad -1 \leqslant x \leqslant 1, \qquad (1)$$

then

$$\omega(f, h) \leqslant Mh \sum_{1 \leqslant n \leqslant h^{-1}} \omega\left(\frac{1}{n}\right). \qquad (2)$$

Proof.   We shall estimate $f(x_1) - f(x_2)$, where $|x_1 - x_2| \leqslant h$. We have one of two possibilities: both x_1, x_2 belong to the same subinterval $[-1, 0]$, $[0, +1]$, or our difference is a sum of two differences with this property. Thus, it is sufficient to estimate the difference $f(x + t) - f(x)$, $x \geqslant 0$, $0 \leqslant t \leqslant h$, $h \leqslant 1$, $x + t \leqslant 1$. Let $k = 0, 1, \cdots$, be arbitrary. By (1),

$$|f(x + t) - f(x)| \leqslant |P_{2^k}(x + t) - P_{2^k}(x)| + 2\omega(\Delta_{2^k}(x)). \qquad (3)$$

Since

$$|P_n(x + t) - P_n(x)| \leqslant h|P_n'(\xi)|, \qquad x < \xi < x + h,$$

we must find an estimate for $P_{2^k}'(y)$. We have

$$|P_{2^k}'(y)| \leqslant |P_1'(y)| + |P_2'(y) - P_1'(y)| + \cdots + |P_{2^k}'(y) - P_{2^{k-1}}'(y)|.$$

We note that

$$\tfrac{1}{4}\Delta_n(y) \leqslant \Delta_{2n}(y) \leqslant \tfrac{1}{2}\Delta_n(y), \qquad -1 \leqslant y \leqslant 1, \qquad n = 1, 2, \cdots. \qquad (4)$$

From (1), 3.5(3), and the first inequality of (4), we obtain

$$|P_{2^i}(y) - P_{2^{i-1}}(y)| \leqslant 2\omega(\Delta_{2^{i-1}}(y)) \leqslant 8\omega(\Delta_{2^i}(y)).$$

By Theorem 3,

$$|P_{2^i}'(y) - P_{2^{i-1}}'(y)| \leqslant \text{const.}\, \Delta_{2^i}(y)^{-1}\omega(\Delta_{2^i}(y)).$$

Also,

$$|P_1(y) - P_0(y)| \leqslant 2\omega(\Delta_0),$$

hence

$$|P_1'(y)| \leqslant 2\omega(\Delta_0) = 2\omega(\Delta_1).$$

Thus, we obtain

$$| P'_{2^k}(y) | \leqslant \text{const} \left\{ \omega(\varDelta_1) + \sum_{i=1}^{k} \varDelta_{2^i}(y)^{-1}\, \omega(\varDelta_{2^i}(y)) \right\}. \tag{5}$$

The last sum is estimated by means of Lemma 1, Chapter 4, with $\phi(u) = \omega(1/u)$, $u_i = \varDelta_{2^i}(y)^{-1}$, $i = 1, 2, \cdots$. This can be done because of (4) and leads to

$$| P'_{2^k}(\xi) | = O(1) \sum_{1 \leqslant n < H_1} \omega \left(\frac{1}{n} \right), \qquad H_1 = \varDelta_{2^k}(\xi)^{-1}. \tag{6}$$

Since ω is increasing,

$$\omega(t) = O(t) \sum_{n \leqslant t^{-1}} \omega \left(\frac{1}{n} \right) \qquad \text{for} \qquad 0 < t \leqslant 1.$$

From this, (6), and (3),

$$| f(x + t) - f(x) | = O(1) \left\{ \varDelta_{2^k}(x) \sum_{n \leqslant H} \omega \left(\frac{1}{n} \right) + h \sum_{n < H_1} \omega \left(\frac{1}{n} \right) \right\}, \tag{7}$$

where $H = \varDelta_{2^k}(x)^{-1}$.

To obtain (2), it seems natural to select $k = k(x)$ in such a way that $\varDelta_{2^k}(x)$ is approximately h. And this works, if one includes a safety factor 4. Let k be chosen so that $4h \leqslant \varDelta_{2^k}(x) < 16h$ [compare (4); if $4h > 1$, this choice is impossible, but then $k = 0$ will serve trivially]. We show that in this case

$$\varDelta_{2^k}(\xi) \geqslant \tfrac{1}{2} \varDelta_{2^k}(x). \tag{8}$$

In fact, if $\varDelta_{2^k}(x) = 2^{-2k}$, then since $\xi > x > 0$, $\varDelta_{2^k}(\xi) = 2^{-2k}$. If, on the other hand, $\varDelta_{2^k}(x) = 2^{-k} \sqrt{1 - x^2}$, then $\sqrt{1 - x^2} \geqslant 2^{-k}$, and from the definition of k, $4h \leqslant \varDelta_{2^k}(x) = 2^{-k} \sqrt{1 - x^2} \leqslant 1 - x^2$. But then

$$\sqrt{1 - \xi^2} \geqslant \sqrt{1 - (x + h)^2} \geqslant \sqrt{1 - x^2 - 3h} \geqslant \tfrac{1}{2} \sqrt{1 - x^2},$$

and we again obtain (8). With the chosen value of k, $H \leqslant h^{-1}$, $H_1 \leqslant h^{-1}$, and hence from (7),

$$| f(x + t) - f(x) | = O(1) h \sum_{n \leqslant h^{-1}} \omega \left(\frac{1}{n} \right), \qquad 0 \leqslant t \leqslant h. \qquad \blacksquare$$

The theorem concerned with differentiable functions is

THEOREM 5. There are constants M_p, $p = 1, 2, \cdots$, with the following property: If ω is a modulus of continuity for which

$$\sum_{n=1}^{\infty} \frac{1}{n}\, \omega \left(\frac{1}{n} \right) < + \infty, \tag{9}$$

and if for $f \in C[-1, +1]$ and polynomials $P_n(x)$ of degree $n = 0, 1, \cdots$,

$$| f(x) - P_n(x) | \leqslant \varDelta_n(x)^p \, \omega(\varDelta_n(x)), \qquad -1 \leqslant x \leqslant 1, \qquad (10)$$

then f has continuous derivatives $f', \cdots, f^{(p)}$, and

$$| f^{(p)}(x) - P_n^{(p)}(x) | \leqslant M_p \sum_{k \geqslant [\varDelta_n(x)^{-1}]} \frac{1}{k} \, \omega\left(\frac{1}{k}\right), \qquad -1 \leqslant x \leqslant 1. \qquad (11)$$

The *proof* is very similar to that of Theorem 8, Chapter 4, and we can omit the details. We have $f = P_n + \sum_{i=1}^{\infty} (P_{2^i n} - P_{2^{i-1} n})$, and the series converges uniformly because of (10). The differentiated series

$$P_n^{(p)}(x) + \sum_{i=1}^{\infty} \{P_{2^i n}^{(p)}(x) - P_{2^{i-1} n}^{(p)}(x)\} \qquad (12)$$

also converges uniformly (and thus represents $f^{(p)}$), for

$$| P_{2^i n}(x) - P_{2^{i-1} n}(x) | \leqslant 2\varDelta_{2^{i-1} n}(x)^p \, \omega(\varDelta_{2^{i-1} n}(x)).$$

Applying (4) and Theorem 3 to this, we obtain the estimate

$$| P_{2^i n}^{(p)}(x) - P_{2^{i-1} n}^{(p)}(x) | = O(1) \, \omega(\varDelta_{2^i n}(x)).$$

The sum in (12) represents the difference $f^{(p)} - P_n^{(p)}$. It is equal to

$$O(1) \sum_{i=1}^{\infty} \omega(\varDelta_{2^i n}(x)) = O(1) \sum_{\left[\frac{\varDelta_{2n}(x)^{-1}}{2}\right]} \frac{1}{k} \, \omega\left(\frac{1}{k}\right),$$

according to Lemma 1 of Chapter 4. It remains to observe that the last sum tends to zero uniformly in x because

$$[\tfrac{1}{2} \varDelta_{2n}(x)^{-1}] \geqslant [\varDelta_n(x)^{-1}] \geqslant n. \qquad \blacksquare$$

It is clear that, from (11) and Theorem 4, we can obtain estimates of the modulus of continuity of the derivative $f^{(p)}$ [compare Sec. 6, Note 1]. Sometimes, combining this with Theorem 2, we can describe the approximation class completely. We give only the most important special case.

THEOREM 6. Let $0 < \alpha < 1$, $p = 0, 1, \cdots$, and $f \in C[-1, +1]$. A necessary and sufficient condition for the existence of a sequence of polynomials $P_n(x)$ with

$$| f(x) - P_n(x) | \leqslant \varDelta_n(x)^{p+\alpha}, \qquad n = 0, 1, \cdots, \qquad -1 \leqslant x \leqslant 1, \qquad (13)$$

is that $f', \cdots, f^{(p)}$ exist and that $f^{(p)} \in \mathrm{Lip}_{M_p} \alpha$, where the constant M_p depends only on p. The proof is left to the reader.

5. Approximation of Analytic Functions

A real- or complex-valued function f defined on $I = [-1, +1]$ is called *analytic on I* if there exists an analytic extension of f onto some open set G of the complex plane that contains I. We mean by this that there must exist on G a single-valued analytic function that coincides with f on I. If this extension exists, it is unique.

Examples of open sets that contain I are the open elliptic discs D_ρ, $\rho > 1$, bounded by the ellipses E_ρ with foci ± 1 and the sum of the half-axes ρ [see 3.4(2)]. Moreover, for each open set $G \supset I$, we have $D_\rho \subset G$, if $\rho > 1$ is sufficiently close to 1.

From the properties of analytic functions, it follows that for each f analytic on I, there exists a $\rho_0 > 1$ characterized by the property that f has an analytic extension onto the disc D_{ρ_0}, but not onto any of the D_ρ for $\rho > \rho_0$. We must, however, admit the posibility $\rho_0 = +\infty$, which is realized for functions f analytic in the whole plane.

The degree of approximation, $E_n(f)$, of an analytic function f by algebraic polynomials satisfies, according to Theorem 6, $E_n(f) = O(n^{-A})$ for each $A > 0$. Actually, much more is true: $E_n(f)$ tends to zero not slower than a geometric progression q^n, $0 < q < 1$. The following theorem, due to Bernstein, ([35] or [2, vol. I, pp. 41, 93]) even gives an asymptotic formula for $E_n(f)$. This is relation (1), given below. This formula is reminiscent of the relation

$$R^{-1} = \overline{\lim} \sqrt[n]{|c_n|}$$

between the radius R of the largest circle of analyticity of a function $g(z) = \sum_0^\infty c_n z^n$ and the coefficients c_n of its Taylor expansion.

THEOREM 7.   A function f, defined on $[-1, +1]$, is analytic on this interval if and only if $\overline{\lim} \sqrt[n]{E_n(f)} < 1$; and more exactly

$$\overline{\lim_{n \to \infty}} \sqrt[n]{E_n(f)} = \frac{1}{\rho_0}, \tag{1}$$

where ρ_0 is the number defined above.

In particular, f has an analytic extension onto the whole plane if and only if

$$\lim_{n \to \infty} \sqrt[n]{E_n(f)} = 0.$$

Proof.   (a)   First we show that if f is analytic in D_ρ, $1 < \rho \leqslant +\infty$, then

$$\overline{\lim} \sqrt[n]{E_n(f)} \leqslant \rho^{-1}. \tag{2}$$

We begin by expanding f on $[-1, +1]$ into a series of Chebyshev polynomials (Chapter 2, Sec. 7):

$$f(x) = \tfrac{1}{2} a_0 + \sum_{k=1}^{\infty} a_k C_k(x). \tag{3}$$

To obtain this, we note that $f(\cos t)$ is an even, 2π-periodic function with a continuous derivative, and has a convergent Fourier series:

$$f(\cos t) = \tfrac{1}{2} a_0 + \sum_{k=1}^{\infty} a_k \cos kt, \qquad a_k = \pi^{-1} \int_{-\pi}^{\pi} f(\cos t) \cos kt \, dt.$$

Substituting $\cos t = x$, we obtain (3). From (3) we have, since $\mid C_n(x) \mid \leqslant 1$ on $[-1, +1]$,

$$E_n(f) \leqslant \sum_{k=n+1}^{\infty} \mid a_k \mid. \tag{4}$$

We see that we must estimate $\mid a_k \mid$. The substitution $z = e^{it}$ in the integral for a_k gives the line integral along the circle $\mid z \mid = 1$:

$$a_k = \frac{1}{\pi i} \int_C f\left(\frac{z + z^{-1}}{2}\right) \frac{z^k + z^{-k}}{2} \frac{dz}{z}. \tag{5}$$

We take a ρ_1 with $1 < \rho_1 < \rho$. Consider the function

$$g(z) = f\left(\frac{z + z^{-1}}{2}\right)$$

in the closed ring R bounded by the circles $C_1 : \mid z \mid = \rho_1^{-1}$ and $C_2 : \mid z \mid = \rho_1$. We know (Chapter 3, Sec. 4) that if z is on either of the circles $\mid z \mid = \sigma$ or $\mid z \mid = \sigma^{-1}$, then $w = \tfrac{1}{2}(z + z^{-1})$ is on the ellipse E_σ. Since $f(w)$ is analytic in D_ρ, it follows that $g(z)$ is analytic in the ring R. In order to obtain a good estimate of a_k, we now change the path of integration. We split the integral (5) into two parts; the integral containing z^{-k} is taken over a circle with a large radius, and for the integral with z^k, we take a circle with a small radius. Thus,

$$a_k = \frac{1}{2\pi i} \int_{C_1} g(z) z^{k-1} dz + \frac{1}{2\pi i} \int_{C_2} g(z) z^{-(k+1)} dz.$$

Let M be the maximum of the absolute value of $f(w)$ on E_{ρ_1}. Then the absolute value of the first integral is not greater than

$$\frac{1}{2\pi} M \left(\frac{1}{\rho_1}\right)^{k-1} 2\pi \rho_1^{-1} = M \rho_1^{-k}.$$

In the same way, the second integral is majorized by $M\rho_1^{-k}$, and we get $|a_k| \leqslant 2M\rho_1^{-k}$. Then, by (4),

$$E_n(f) \leqslant 2M \sum_{k=n+1}^{\infty} \rho_1^{-k} = \frac{2M}{\rho_1 - 1} \rho_1^{-n} = M_1 \rho_1^{-n}. \qquad (6)$$

Hence, $\sqrt[n]{E_n(f)} \leqslant M_1^{1/n}\rho_1^{-1}$, and we obtain $\overline{\lim}\ \sqrt[n]{E_n(f)} \leqslant \rho_1^{-1}$. This gives (2), since ρ_1 can be taken arbitrarily close to ρ.

(b) Conversely, let (2) hold for some ρ, $1 < \rho \leqslant +\infty$. We show that f has an analytic extension onto each elliptic disc D_{ρ_1}, $1 < \rho_1 < \rho$ (and therefore onto D_ρ). Let $\rho_1 < \rho_2 < \rho$; then $\sqrt[n]{E_n(f)} \leqslant \rho_2^{-1}$ for $n \geqslant n_0$ with a sufficiently large n_0 . Writing

$$f(x) = P_0(x) + \sum_{n=0}^{\infty} [P_{n+1}(x) - P_n(x)], \qquad (7)$$

where the P_n are the polynomials of best approximation for f, we have

$$|P_{n+1}(x) - P_n(x)| \leqslant 2E_n(f) \leqslant 2\rho_2^{-n}, \qquad -1 \leqslant x \leqslant 1, \qquad n \geqslant n_0 .$$

Using the inequality 3.4(3), we can estimate this difference for all z in D_{ρ_1}:

$$|P_{n+1}(z) - P_n(z)| \leqslant 2\rho_1 \left(\frac{\rho_1}{\rho_2}\right)^n, \qquad z \in D_{\rho_1}, \qquad n \geqslant n_0 .$$

Hence, the series (7), with x replaced by z, converges uniformly on each compact subset of D_ρ. Its sum is analytic on D_ρ and provides the desired analytic extension of f.

It follows from (a) and (b) that the largest elliptic disc D_ρ in which f is analytic is D_{ρ_0} , with ρ_0 given by (1). The proof is complete. ∎

6. Notes

1. We can reformulate the conclusions (2) and (11) of Theorems 4 and 5 in integral form:

$$\omega(f, h) \leqslant Mh \int_h^1 \frac{\omega(x)}{x^2}\, dx, \qquad h \geqslant 0, \qquad (1)$$

$$|f^{(p)}(x) - P_n^{(p)}(x)| \leqslant M_p \Omega(\varDelta_n(x)), \qquad \Omega(h) = \int_0^h \frac{\omega(x)}{x}\, dx. \qquad (2)$$

(It is easy to see that Ω is a modulus of continuity whenever ω is.) Applying (1) to (2) and changing the order of integration, we obtain an estimate of the modulus of continuity of $f^{(p)}$, if the conditions of Theorem 5 are satisfied:

$$\omega(f^{(p)}, h) \leqslant M_p \left\{ \int_0^h \frac{\omega(x)}{x}\, dx + h \int_h^1 \frac{\omega(x)}{x^2}\, dx \right\}.$$

(Compare [15, p. 359].)

2. In the formulation of Theorem 1, it is possible to replace $\omega(f, h)$ by $\omega_p(f, h)$, $p = 2, 3, \cdots$ (Dzjadyk [44], Freud [51], Brudnyĭ [37]). While in the proof of Theorem 2, Chapter 4, it does not require any additional work to replace ω by ω_2, the situation is different for the approximation by algebraic polynomials. The reason is that the second difference of the function $f(\cos t)$ is not trivially expressible as a second difference of $f(x)$.

The corresponding inverse theorems can also be obtained. We mention a simple special case. A function $f \in C[-1, +1]$ is quasi-smooth if and only if, for a sequence of polynomials P_n, $|f(x) - P_n(x)| \leqslant M\Delta_n(x)$.

3. For polynomials in x, $1 - x$ with positive coefficients

$$P_n(x) = \sum_{k+l \leqslant n} a_{kl} x^k (1 - x)^l, \qquad a_{kl} \geqslant 0 \tag{3}$$

and the interval $[0, 1]$, similar results hold. Of course only positive functions $f \in C[0, 1]$ can be approximated. Roughly, the results of the present chapter remain valid for the polynomials (3) if one replaces n by $\sqrt{n}$ in the estimates of the degree of approximation. For instance, $f \in \Lambda_{p\alpha}$ holds if and only if for a sequence of polynomials (3), $|f(x) - P_n(x)| \leqslant \text{const}\, \overline{\Delta}_n(x)^{p+\alpha}$, where $\overline{\Delta}_n(x) = \max(\sqrt{x(1 - x)}\, n^{-1}, n^{-1})$. See Lorentz [75].

PROBLEMS

1. For functions f with a continuous derivative, $E_n(f) \leqslant Mn^{-1}E_{n-1}(f')$.

2. Derive from the argument of page 65 and Problem 1 that, for each $p = 1, 2, \cdots$, there is a constant M_p for which

$$E_n(f) \leqslant M_p n^{-p} \omega \left(f^{(p)}, \frac{1}{n} \right), \qquad n > p,$$

if $f^{(p)} \in C[a, b]$.

3. If f, $f^{(p)} \in C[a, b]$ and $f^{(s)}(x) \geqslant 0$, $a \leqslant x \leqslant b$, for some s with $0 \leqslant s < p$, then there is a sequence of polynomials $P_n(x)$ for which $P_n^{(s)}(x) \geqslant 0$ and

$$\| f - P_n \| \leqslant \text{const}\, n^{s-p} \omega \left(f^{(p)}, \frac{1}{n} \right) \qquad \text{(O. Shisha).}$$

4. If $f \in C[0, 1]$, $f(x) \geqslant 0$ for $0 \leqslant x \leqslant 1$, then there is a ppc P_n with

$$|f(x) - P_n(x)| \leqslant 2\omega(f, \delta_n), \qquad \delta_n = \sqrt{x(1 - x)/n} \qquad (*)$$

(see Chapter 3, Sec. 9, Note 2).

5. If, in addition, $f'(x)$ exists, then (*) can be replaced by

$$|f(x) - P_n(x)| \leqslant 2\delta_n \omega(f', \delta_n).$$

(*Hint*: Use Bernstein polynomials in Problems 4 and 5.)

6. If $Q(x) = ax^2 + bx + c$, then $B_n(Q, x) - Q(x) = an^{-1}x(1 - x)$.

7. If $f \in C[0, 1]$, $f(\alpha) = f(\beta) = 0$ for some $0 \leqslant \alpha < \beta \leqslant 1$, and if f has a positive maximum on $[\alpha, \beta]$, then there exists a polynomial $Q(x) = ax^2 + bx + c$, $a < 0$, for which $f(x) \leqslant Q(x)$, $\alpha \leqslant x \leqslant \beta$ and $f(\gamma) = Q(\gamma)$ for some γ, $\alpha < \gamma < \beta$.

8. If for a function $f \in C[0, 1]$, $f(x) - B_n(x) = o(1/n)$ for each $x \in [\alpha, \beta]$, then $f(x)$ is linear on $[\alpha, \beta]$ (Bajsanski and Bojanic). [*Hint*: Assume $f(\alpha) = f(\beta) = 0$. Then $f(x) \leqslant Q(x) + g(x)$, where Q is as in Problem 7, and g is a bounded function which vanishes on $[\alpha, \beta]$.]

9. Prove (1), (2) in Note 1.

[6]

Approximation by Rational Functions.
Functions of Several Variables

1. Degree of Rational Approximation

The main subject of this book is *linear approximation of functions*, that is, approximation by linear combinations of given functions. As an important case of *nonlinear approximation*, we shall discuss approximation by rational functions. A rational function $r_n(x)$ of degree n is the quotient of two polynomials of degree n without common zeros:

$$r_n(x) = \frac{P_n(x)}{Q_n(x)}, \qquad -\infty < x < +\infty. \tag{1}$$

First some simple remarks about rational functions. The representation (1) is unique, up to constant factors of P_n and Q_n. A rational function $r_n(x)$ may take infinite values for some real x. The sum of two rational functions of degrees n and m is a rational function of degree $n + m$.

Each function $f \in C[a, b]$ is approximable by rational functions; this is an immediate corollary of the theorem of Weierstrass. Theorems which describe rational functions of best approximation are also known. Let $f \in C[a, b]$. It is known that the rational function r_n of degree n of best approximation is unique (Walsh [18, p. 351]); the difference $f - r_n$ has the Chebyshev alternation property on $[a, b]$ (Ahiezer [1, p. 55]). An exposition of results of this type in a general setting will be found in Cheney [38].

Leaving these problems aside, we shall restrict ourselves, in Secs. 1 and 2, to a study of the asymptotic behavior of $R_n(f)$, the nth *degree of approximation of f by rational functions*. If f is a continuous function on a subset A of the real line, we define

$$R_n(f) = R_n(f, A) = \inf_{r_n} \{\sup_{x \in A} |f(x) - r_n(x)|\}, \tag{2}$$

where the infimum is taken for all r_n that are finite on A.

Clearly, $R_n(f) \leqslant E_n(f)$, where $E_n(f)$ is the nth degree of approximation of f by polynomials on A.

How much smaller can $R_n(f)$ be than $E_n(f)$? Results of two kinds are known in this direction. On one hand, for many natural classes of functions $\mathfrak{R}$, the suprema $\sup_{f \in \mathfrak{R}} R_n(f)$ and $\sup_{f \in \mathfrak{R}} E_n(f)$ are asymptotically the same. This shows that for "all but the best" functions of $\mathfrak{R}$, it is not worth while to use rational approximation. Theorems of this type have been obtained by Vituškin [17] and Erohin [47]. Also, Theorems 3 and 4 of this chapter point in the same direction.

On the other hand, for functions with only few singularities, rational approximation may be very useful. For example, consider $f(x) = \Gamma(x + 2)$ on $[-1, +1]$. The singular point of this function closest to $[-1, +1]$ is the pole $x = -2$ with the singular part $(x + 2)^{-1}$. Hence, by Theorem 8, Chapter 5, $E_n(f)$ is of the order of ρ^{-n}, where $\rho = 2 + \sqrt{3}$ is given by $\frac{1}{2}(\rho + \rho^{-1}) = 2$. The next pole of f is at $x = -3$. By the same theorem, for certain polynomials P_n, the difference $f(x) - \{(x + 2)^{-1} + P_n(x)\}$ is on $[-1, +1]$ of the order ρ_1^{-n}, $\rho_1 = 3 + \sqrt{8}$.

Only very recently natural classes of continuous functions f have been found, for which $R_n(f)$ is of different order of magnitude than $E_n(f)$. See Note 4. We will be content to discuss in detail a very special, but important function. Let $h(x) = |x|$ on $A = [-1, +1]$. In 7.1(5) we shall see that $E_n(h) \approx n^{-1}$. In contrast to this:

THEOREM 1 (D. J. NEWMAN [82]).   We have

$$R_n(h) \leqslant 3e^{-\sqrt{n}}, \qquad n \geqslant 5. \tag{3}$$

Proof. Let $a = e^{-1/\sqrt{n}}$; if n is large, a is close to 1, but a^{n-1} is close to 0. First we show that

$$\prod_{j=1}^{n-1} \frac{1 - a^j}{1 + a^j} \leqslant e^{-\sqrt{n}}, \qquad n \geqslant 5. \tag{4}$$

The inequality $(1 - t)/(1 + t) \leqslant e^{-2t}$ is true for $t \geqslant 0$; therefore,

$$\prod_{j=1}^{n-1} \frac{1 - a^j}{1 + a^j} \leqslant \exp\left\{-2 \sum_{j=1}^{n-1} a^j\right\} = \exp\left\{-2 \frac{a - a^n}{1 - a}\right\}.$$

But

$$2(a - a^n) \geqslant 2(e^{-1/\sqrt{5}} - e^{-\sqrt{5}}) > 1 \qquad \text{for} \qquad n \geqslant 5.$$

Also, for all $t \geqslant 0$, $1 - e^{-t} \leqslant t$; hence, $1/(1 - a) \geqslant \sqrt{n}$, and (4) follows.

Now let

$$p(x) = \prod_{k=1}^{n-1} (x + a^k), \qquad r_n(x) = x \frac{p(x) - p(-x)}{p(x) + p(-x)}. \tag{5}$$

Clearly, r_n is a rational function of degree n. We shall show that $||x| - r_n(x)| < 3e^{-\sqrt{n}}$, $n \geqslant 5$, $x \in A$. Since $|x|$ and $r_n(x)$ are both even, it suffices to consider the case when $0 \leqslant x \leqslant 1$. For $0 \leqslant x \leqslant a^n = \exp(-\sqrt{n})$, this is quite trivial, since here $p(-x) \geqslant 0$, so that $0 \leqslant r_n(x) \leqslant x$. Hence, $||x| - r_n(x)| \leqslant x \leqslant e^{-\sqrt{n}}$.

Next we have

$$|\,|x\,| - r_n(x)\,| = 2x\left|\frac{p(-x)}{p(x) + p(-x)}\right| \leqslant \frac{2}{\left|\dfrac{p(x)}{p(-x)}\right| - 1}, \qquad 0 \leqslant x \leqslant 1. \quad (6)$$

If $a^n \leqslant x \leqslant 1$, then for some j, $0 \leqslant j \leqslant n - 1$, $a^{j+1} \leqslant x \leqslant a^j$. In this case,

$$\left|\frac{p(-x)}{p(x)}\right| = \prod_{k=1}^{j}\frac{a^k - x}{a^k + x}\prod_{k=j+1}^{n-1}\frac{x - a^k}{x + a^k} \leqslant \prod_{k=1}^{j}\frac{a^k - a^n}{a^k + a^n}\prod_{k=j+1}^{n-1}\frac{a^j - a^k}{a^j + a^k}$$

$$= \prod_{m=n-j}^{n-1}\frac{1 - a^m}{1 + a^m}\prod_{m=1}^{n-j-1}\frac{1 - a^m}{1 + a^m} = \prod_{m=1}^{n-1}\frac{1 - a^m}{1 + a^m} \leqslant e^{-\sqrt{n}},$$

by (4). Hence, by (6), if $a^n \leqslant x \leqslant 1$,

$$|\,|x\,| - r_n(x)\,| \leqslant \frac{2}{e^{\sqrt{n}} - 1} \leqslant 3e^{-\sqrt{n}}, \qquad \text{if} \qquad n \geqslant 5. \qquad \blacksquare$$

The reader will find in [82] the proof that $R_n(h) \geqslant \frac{1}{2}e^{-9\sqrt{n}}$, $n \geqslant 5$.

2. Further Theorems

Direct theorems of rational approximation follow from results of Chapter 5, and the inequality $R_n(f) \leqslant E_n(f)$. For example, if $f \in \Lambda_{p\alpha}$, $p = 0, 1, \cdots$, $0 < \alpha \leqslant 1$ on $[a, b]$, then $R_n(f) = O(n^{-p-\alpha})$. Inverse theorems that we shall obtain in this section do not quite match these; their formulations contain a small exceptional set. If $R_n(f, A)$ is very small, we can deduce only that f is very smooth on $A \setminus e$, where e is a set of arbitrary small measure. Theorem 1 shows that this is necessary.

In the same way, if $|\,r_n(x)\,| \leqslant M$ on $[a, b]$, there may be points inside $[a, b]$ where r'_n is large. For example, $r(x) = \delta^2/(x^2 + \delta^2)$, where δ is small, satisfies $|\,r(x)\,| \leqslant 1$ on $[-1, +1]$, but $r'(\delta) = -\frac{1}{2}\delta^{-1}$ is large. Thus, Bernstein's inequality does not hold for rational functions. We can derive, however, a modified form of this inequality.

THEOREM 2. Let r_n be a rational function of degree n, let $A \subset (-\infty, +\infty)$, and let $|\,r_n(x)\,| \leqslant M$ on A. Then, for each $\delta > 0$, there is a subset e of $(-\infty, +\infty)$, $me \leqslant \delta$, such that on $A \setminus e$, r_n belongs to the class Lip 1 with the constant $(2/\delta)Mn$:

$$|\,r_n(x) - r_n(x')\,| < \frac{2Mn}{\delta}\,|\,x - x'\,|, \qquad x, x' \in A \setminus e. \quad (1)$$

Proof. We can assume that r_n is not a constant. Let F be the subset of $(-\infty, +\infty)$ where $|\,r_n(x)\,| \leqslant M$. Then $F \supset A$. The set F is the union of finitely many intervals I, disjoint except for the end points, on each of which r_n is monotone. Let I' be the range of r_n on I. Then $I' \subset [-M, M]$. Since each equation $r_n(x) = \text{const}$ has at most n solutions, the intervals I' cover $[-M, M]$ at most n times. It follows that

$$\operatorname{Var}_F r_n = \sum mI' \leqslant 2nM.$$

Now let $x_1 < x_1'$ be two points of F, for which

$$| r_n(x_1) - r_n(x_1') | \geqslant \frac{2Mn}{\delta} (x_1' - x_1). \tag{2}$$

This implies $x_1' - x_1 \leqslant (\delta/2Mn) \operatorname{Var}_F r_n \leqslant \delta$. Hence, there is an interval (x_1, x_1') of *maximal length* which satisfies (2). Next let $F_1 = F \setminus (x_1, x_1')$ and let (x_2, x_2'), x_2, $x_2' \in F_1$ be an interval of maximal length, different from (x_1, x_1'), with $| r_n(x_2) - r_n(x_2') | \geqslant 2Mn(x_2' - x_2)/\delta$. This interval is disjoint with (x_1, x_1'), and as before, $(x_1' - x_1) + (x_2' - x_2) \leqslant \delta$. Continuing this construction by induction, we obtain a set e, of measure $me \leqslant \delta$, which is a union of finitely or countably many disjoint intervals (x_n, x_n') and $x_n' - x_n \to 0$ in the last case. On $F \setminus e$, we have (1). ∎

COROLLARIES (Dolženko [41]). 1. In the assumptions of Theorem 2, there is a set e_1, $me_1 \leqslant \delta$ for which

$$| r_n'(x) | \leqslant \frac{2}{\delta} Mn, \qquad x \in A \setminus e_1.$$

2. Under the same assumptions, for each $p = 1, 2, \cdots$,

$$| r_n^{(p)}(x) | \leqslant \frac{C_p M}{\delta^p} n^p, \qquad x \in A \setminus e_p, \tag{4}$$

for some set e_p with measure $me_p \leqslant \delta$; C_p are absolute constants.

Indeed, for e_1 we can take the union of e and the set of all isolated points of $A \setminus e$. Corollary 2 follows from 1 by induction.

Of interest are natural classes of functions, where rational approximation is essentially better than approximation by algebraic polynomials. Some cases of this type are given below.

In the following theorem, V_α is the intersection of Lip α with the space V of functions of bounded variation.

THEOREM 3 (Freud). For a function $f \in V_\alpha$, $0 < \alpha < 1$,

$$R_n(f) \leqslant \text{const.} \frac{\log^2 n}{n} \qquad (5)$$

This will follow from a more general theorem below. Let $\omega(h)$ be a strictly increasing continuous function defined for $h \geqslant 0$, with $\omega(0) = 0$, and let

$$\phi(h) = \log^2(1/h)/\omega(h), \qquad 0 < h < 1.$$

The function $\phi(h)$ strictly decreases on $[0,1]$ and satisfies $\phi(1) = 0$, $\phi(0+) = \infty$. Let $V > 0$ be given. For each n there exists a unique δ_n for which

$$n = 4V \frac{\log^2(1/\delta_n)}{\omega(\delta_n)} \qquad (6)$$

THEOREM 4. Let f be of bounded variation $\text{Var } f = V$ on $[0,1]$, and let $\omega(f,h) \leqslant \omega(h)$. Then

$$R_n(f) \leqslant 18V\delta_n + 2\omega(\delta_n). \qquad (7)$$

Proof. We construct inductively points $c_{0'} = 0 < c_1 < \cdots < c_{s+1} = 1$ in the following manner. Let c_1 be the largest number $c_1 \leqslant 1$ satisfying

$$|f(x) - f(c_0)| < \omega(\delta_n) \qquad \text{for all} \qquad x \in [c_0, c_1].$$

Then either $c_1 = 1$ or $c_1 < 1$, and $|f(c_1) - f(c_0)| = \omega(\delta_n)$. We continue this process; if $0 = c_0 < \cdots < c_k < 1$ are known, then c_{k+1} is the largest number, $c_{k+1} \leqslant 1$ for which

$$|f(x) - f(c_k)| \leqslant \omega(\delta_n) \qquad \text{for} \qquad x \in [c_k, c_{k+1}]. \qquad (8)$$

We have

$$|f(c_{k+1}) - f(c_k)| \leqslant \omega(\delta_n), \qquad \text{with equality if } c_{k+1} < 1. \qquad (9)$$

This process must terminate with some $c_{s+1} = 1$. Since

$$V \geqslant \sum_{k=0}^{s-1} |f(c_{k+1}) - f(c_k)| = s\omega(\delta_n),$$

we have

$$s \leqslant \frac{V}{\omega(\delta_n)}. \tag{10}$$

Moreover, since $|f(x+t) - f(x)| \leqslant \omega(\delta_n)$ for $0 \leqslant t \leqslant \delta_n$, $x + t \leqslant 1$, we have, by the definition of c_{k+1}, $c_{k+1} - c_k \geqslant \delta_n$, $k = 0, \ldots, s-1$. We approximate the function f by a broken linear function g, which interpolates f at the points $0 = c_0 < \cdots < c_{s-1}$ and at $c_{s+1} = 1$. Then

$$|f(x) - g(x)| \leqslant 2\omega(\delta_n), \qquad 0 \leqslant x \leqslant 1. \tag{11}$$

If l_k is the slope of g in the interval beginning with c_k, $k = 0, \ldots, s - 1$, then

$$|l_k| = \frac{|f(c_{k+1}) - f(c_k)|}{c_{k+1} - c_k} \leqslant \frac{\omega(\delta_n)}{\delta_n}, k = 0, \ldots, s - 2$$

$$|l_{s-1}| \leqslant \frac{2\omega(\delta_n)}{\delta_n} \tag{12}$$

The function $g(x)$ can be written in the form

$$g(x) = f(0) + l_0 x + \sum_{k=1}^{s-1} (l_k - l_{k-1})(x - c_k)_+$$

$$= Ax + B + \sum_{k=1}^{s-1} \frac{l_k - l_{k-1}}{2} |x - c_k|.$$

For the integer $m = [2 \log(1/\delta_n)]^2$, let $r_m(x)$ be Newman's approximation of $|x|$ on $[-1, +1]$. Then

$$R(x) = Ax + B + \sum_{k=1}^{s-1} \frac{l_k - l_{k-1}}{2} r_m(x - c_k)$$

is a rational function of degree

$$\leqslant sm \leqslant \frac{V}{\omega(\delta_n)} 4 \log^2(1/\delta_n) = n$$

and since $e^{-\sqrt{m}} \leqslant e \, \delta_n^2$, we have by Theorem 1 and (12),

$$|g(x) - R(x)| \leqslant s \frac{2\omega(\delta_n)}{\delta_n} 3 e^{-\sqrt{m}} \leqslant 6eV\delta_n < 18V\delta_n. \tag{13}$$

Combining (11) and (13), we obtain the theorem.

To derive Theorem 3, we note that in case $f \in \text{Lip } \alpha$, we can take $\omega(h) = Mh^\alpha$. From equation (7) we obtain, for all large n,

$$\delta_n < C(n^{-1} \log^2 n)^{1/\alpha},$$

if $C > 0$ is sufficiently large. Then (6) becomes (5).   ∎

It is not known whether the inequality (5) in Theorem 3 is the best possible.

A very similar calculation yields

THEOREM 5. If f is a function of bounded variation and for some $\beta > 0$ satisfies

$$\omega(f,h) \leqslant M \log^{-\beta} \frac{1}{h}, \qquad h > 0,$$

then $R_n(f) \leqslant \text{Const. } n^{-\frac{\beta}{\beta+2}}$.

3. Periodic Functions of Several Variables

Approximation of functions of several variables is important for purposes of practical computation and for theoretical considerations. We shall need the theorems of this section in Chapter 8.

We shall consider continuous functions $f(x_1, \cdots, x_s)$, on the *s-dimensional orus* $K^{(s)} = K \times \cdots \times K$ (see Chapter 3, Sec. 5).

The norm $\|f\|$ is the maximum of the absolute value of f on $K^{(s)}$. We denote by $T_{n_1 \ldots n_s}(x_1, \cdots, x_s)$, a trigonometric polynomial in the x_i, of degree n_i in x_i, $i = 1, \cdots, s$. The minimum of $\|f - T_{n_1 \ldots n_s}\|$ for all polynomials of given degrees n_i is the degree of approximation $E^*_{n_1 \ldots n_s}(f)$. In the following theorem, the partial moduli of continuity (Chapter 3, Sec. 5) are used. This theorem is a generalization of Theorem 3, Chapter 4, to functions of s variables.

THEOREM 6. Let $f \in C[K^{(s)}]$ be a function that has continuous partial derivatives $\partial^k f / \partial x_i^k$, $0 \leqslant k \leqslant p_i$, $i = 1, \cdots, s$ $(p_i \geqslant 0)$. Assume that the modulus of continuity of $\partial^{p_i} f / \partial x_i^{p_i}$ with respect to x_i does not exceed $\omega_i(h)$, $i = 1, \cdots, s$. Then

$$E^*_{n_1 \ldots n_s}(f) \leqslant M \sum_{i=1}^{s} \frac{1}{n_i^{p_i}} \omega_i \left(\frac{1}{n_i}\right), \qquad n_i \geqslant 1, \qquad i = 1, \cdots, s, \qquad (1)$$

where M is a constant that depends only on the p_i, $i = 1, \cdots, s$.

Proof.   The idea of the proof is gradually to approximate f by a trigonometric polynomial, in each variable in turn. This will be done by means of the operator 4.3(7).

We define some subclasses of $C[K^{(s)}]$. By $\mathfrak{T}_j$, $j = 1, \cdots, s + 1$, we denote the set of all functions $f \in C[K^{(s)}]$ with the following properties: In x_i, $i < j$, f is a trigonometric polynomial of degree n_i; in x_i, $i \geqslant j$, f is a function with continuous partial derivatives $\partial^k f / \partial x_i^k$, $0 \leqslant k \leqslant p_i$; the modulus of continuity of the p_ith derivative with respect to x_i should not exceed $2^{P_j}\omega_i(h)$, where

$$P_j = (p_1 + 1) + \cdots + (p_{j-1} + 1), \qquad j > 1, \qquad P_1 = 1.$$

In particular, $\mathfrak{T}_1$ is the class of all functions f of Theorem 6, and $\mathfrak{T}_{s+1}$ consists of all trigonometric polynomials $T_{n_1,\ldots,n_s}$.

LEMMA 2.   If $f_j \in \mathfrak{T}_j$, $j = 1, \cdots, s$, then there exists a function $f_{j+1} \in \mathfrak{T}_{j+1}$, for which

$$\| f_j - f_{j+1} \| \leqslant M \frac{1}{n_j^{p_j}} \omega_j \left(\frac{1}{n_j} \right). \tag{2}$$

We shall find f_{j+1} in the form $f_{j+1} = I_{n_j r_j}(f_j, x_j)$, where the operator 4.3(7) acts on f_j, regarded as a function of x_j. More precisely,

$$f_{j+1}(x_1, \cdots, x_s) = - \int_{-\pi}^{\pi} K_{n_j r_j}(t) \sum_{k=1}^{p_j+1} (-1)^k \binom{p_j + 1}{k}$$

$$\times f_j(x_1, \cdots, x_j + kt, \cdots, x_s) \, dt, \tag{3}$$

where r_j is the smallest integer for which $r_j \geqslant \frac{1}{2}(p_j + 3)$. Clearly, f_{j+1}, together with f_j, is a trigonometric polynomial in x_i, of degree n_i, $i < j$. In x_j, f_{j+1} is a trigonometric polynomial of degree n_j, according to the proof of Theorem 3, Chapter 4. Finally, we can differentiate (3) with respect to x_i, $i > j$, under the integration sign, and obtain

$$\frac{\partial^{p_i} f_{j+1}}{\partial x_i^{p_i}} = I_{n_j p_j} \left(\frac{\partial^{p_i} f_j}{\partial x_i^{p_i}}, x_j \right). \tag{4}$$

Since the kernel $K_{n_j r_j}$ is positive and has an integral equal to 1 over $[-\pi, \pi]$, we derive from (4) that the partial modulus of continuity of $\partial^{p_i} f_{j+1} / \partial x_i^{p_i}$ with respect to x_i does not exceed

$$\sum_{k=1}^{p_j+1} \binom{p_j + 1}{k} \omega_i \left(\frac{\partial^{p_i} f_j}{\partial x_i^{p_i}}, h \right) \leqslant 2^{p_j+1} 2^{P_j} \omega_i(h) = 2^{P_{j+1}} \omega_i(h).$$

Thus, $f_{j+1} \in \mathfrak{T}_{j+1}$. According to 4.3(10), relation (2) holds with $M = M'_{p_j}$. ∎

From (2) we obtain by iteration, starting with $f_1 = f$,

$$\| f - f_{s+1} \| \leqslant M \sum_{i=1}^{s} \frac{1}{n_i^{p_i}} \, \omega_i \left(\frac{1}{n_i} \right). \tag{5}$$

As a corollary we have

THEOREM 7.   For each function $f \in \Lambda_{p\omega}^s[K^{(s)}]$, there is a trigonometric polynomial $T_n(x_1, \cdots, x_s)$ of degree n in each variable such that

$$\| f - T_n \| \leqslant M \frac{1}{n^p} \, \omega \left(\frac{1}{n} \right). \tag{6}$$

4.   Approximation by Algebraic Polynomials

By means of Theorem 6, we can obtain results concerning approximation of functions of $s = 2, 3, \cdots$ variables $f(x_1, \cdots, x_s)$, defined on the cube B: $-1 \leqslant x_i \leqslant 1$, $i = 1, \cdots, s$ by algebraic polynomials. We discuss only the uniform approximation, and do not consider the possibility of better approximation toward the boundary of B. The following lemma deals with functions of one variable.

LEMMA 3.   For each class $\Lambda = \Lambda_{p\omega}(M_0, \cdots, M_{p+1}; [-1, +1])$, there exists a constant C with the following property: If $f \in \Lambda$, then the function $g(t) = f(\cos t)$ is p-times differentiable and

$$\omega(g^{(p)}, h) \leqslant C\omega(h), \qquad 0 < h \leqslant 1. \tag{1}$$

Proof.   By induction we obtain

$$g^{(k)}(t) = \sum_{j=1}^{k} f^{(j)}(\cos t) \, T_{kj}(t), \qquad k = 1, \cdots, p, \tag{2}$$

where the T_{kj} are some trigonometric polynomials, independent of f. We select $M' > 0$ so that $\| T_{kj} \| \leqslant M'$, $\| T'_{kj} \| \leqslant M'$ for all k, j; then $\omega(T_{kj}, h) \leqslant M'h$. The modulus of continuity of $f^{(j)}(\cos t)$ does not exceed $\| f^{(j+1)} \| \, h \leqslant M_{j+1}h$ if $j < p$, and $M_{p+1}\omega(h)$ if $j = p$. We apply 3.5(5) to each term of (2) (with $k = p$):

$$\omega(g^{(p)}, h) \leqslant M' \left\{ \sum_{j=1}^{p-1} (M_{j+1}h + M_jh) + M_{p+1}\omega(h) + M_ph \right\}$$

$$\leqslant C'h + M'M_{p+1}\omega(h). \tag{3}$$

But for $0 < h \leqslant 1$, $\omega(1) \leqslant (h^{-1} + 1) \, \omega(h) \leqslant 2h^{-1}\omega(h)$, so that $h \leqslant 2\omega(h)/\omega(1)$. Therefore, (1) follows from (3).

THÈOREM 8. If a class $\varLambda = \varLambda_p^s \, (M_0, \cdots, M_{p+1}; \, B)$ is given, there exists a constant M with the property that for each function $f \in \varLambda$, there are polynomials $P_{n_1 \ldots n_s}(x_1, \cdots, x_s)$, of degree n_i in x_i, for which

$$\| f - P_{n_1 \ldots n_s} \| \leqslant M \sum_{i=1}^{s} \frac{1}{n_i^p} \, \omega \left(\frac{1}{n_i} \right), \qquad n_i = 1, 2, \cdots, \qquad i = 1, \cdots, s. \tag{4}$$

Proof. The function

$$g(t_1, \cdots, t_s) = f(\cos t_1, \cdots, \cos t_s) \tag{5}$$

is defined on $K^{(s)}$. By Lemma 3, the moduli of continuity of the derivatives $\partial^p g / \partial t_i^p$ do not exceed $C\omega(h)$.

Then, by Theorem 6, there are trigonometric polynomials $T_{n_1 \ldots n_s}$ for which

$$\| g - T_{n_1 \ldots n_s} \| \leqslant M \sum_{i=1}^{s} \frac{1}{n_i^p} \, \omega \left(\frac{1}{n_i} \right).$$

Since the function g is even in each variable, we can assume that $T_{n_1 \ldots n_s}$ are pure cosine polynomials. With the substitution $\cos t_i = x_i$, we obtain a polynomial $P_{n_1 \ldots n_s}$, which satisfies (4).   ∎

5. Notes

1. In [41], Dolženko gives many variations of the inequalities 2(3) and 2(4), in the complex plane and for rational functions of several variables.

2. Theorem 5 is slightly stronger than the results of Gončar [54]; he proves that $R_n(f, A) = O(n^{-p-\alpha-\epsilon})$, $\epsilon > 0$ implies $f^{(p)} \in \mathrm{Lip}_{C'} \alpha$ on $A \setminus e$, $me \leqslant \delta$ with some unspecified $C'(\delta)$.

3. The estimates of Theorems 6, 7, and 8 become meaningless if both n and s are very large. The only known useful result in this situation is due to Newman and Shapiro [83].

4. Newman's inequality 1(3) can be used to estimate the degree of approximation $R_n(f)$ by rational functions of many special functions f. In this way Szüsz and Turán, and later Freud, have discovered natural classes of functions for which $R_n(f)$ is of different order than $E_n(f)$. Freud ["Über die Approximation reeller Funktionen durch rationale gebrochene Funktionen," to appear in *Acta Math. Acad. Hungar.*, vol. **17** (1966)] proves, for example, that $R_n(f) \leqslant$ const $Mn^{-1} \log^2 n$ if $f \in \mathrm{Lip}_M \alpha$ and $\mathrm{Var} f \leqslant M$, and that $R_n(f) \leqslant$ const $Mn^{-\gamma/(\gamma+2)}$ if $|f(x + t) - f(x)| \leqslant M \log^{-\gamma}(1/|t|)$ and $\mathrm{Var} f \leqslant M$.

PROBLEMS

1. In the situation of Theorem 4, f' exists with respect to A almost everywhere on A.

2. Refining 2(4) somewhat, show the existence of a set $e \subset A$, $me \leqslant \delta$ for which $|r_n^{(p)}(x)| \leqslant p!\, 2^{p(p+3)/2}(n/\delta)^p M$, $x \in A \setminus e$, for each $p = 1, 2, \cdots$ (Dolženko).

3. Let G be the set of the complex plane where $|r_n(z)| \leqslant M$; let $\delta > 0$. Then there exists a subset e of G, $me \leqslant \delta$, such that $|r_n(z)| \leqslant M\sqrt{\pi n/\delta}$ on $G \setminus e$ (Dolženko). [*Hint:* The area of the image of G is $\int_G |r_n'(z)|^2\, dz$].

4. Assume that $\phi_n > 0$, $\delta_n > 0$, $\sum \phi_n < +\infty$. Then, for the function

$$f(x) = \sum_{k=0}^{\infty} \phi_k \delta_k (1 + \delta_k - x)^{-1}, \qquad 0 \leqslant x \leqslant 1,$$

$R_n(f) \leqslant \sum\limits_{k=n}^{\infty} \phi_k$, while $\omega(f, \delta_k) \geqslant \frac{1}{2}\phi_k$.

5. Let $\phi_n \to 0$, $\rho_n \to 0$, $\phi_n > 0$, $\rho_n > 0$ be arbitrary. There exists a function $f \in C[0, 1]$, for which $R_n(f) \leqslant \phi_n$, $n = 1, 2, \cdots$, while $E_n(f) \geqslant \rho_n$ for an infinity of n. [(Gončar) *Hint:* Use Problem 4 and Theorem 4, Chapter 5.]

6. Let $E_{n\infty}^*(f)$ be the degree of approximation of $f \in C[K^{(2)}]$ by continuous functions $g(x, y)$, which are trigonometric polynomials of degree n in x, and let $E_{\infty m}^*$ be defined similarly. Using Theorem 1 of Chapter 4, prove that

$$E_{nm}^*(f) \leqslant M\{E_{n\infty}^*(f) + E_{\infty m}^*(f)\} \log n \qquad n > 1 \qquad \text{(Bernstein)}.$$

[7]

Approximation
by Linear Polynomial Operators

1. Sums of de la Vallée-Poussin. Positive Operators

A bounded linear operator L_n that maps $C[-1, +1]$ into itself is a *polynomial operator of degree n* if each of its values $L_n(f)$, $f \in C[-1, +1]$, is an algebraic polynomial of degree n. Trigonometric polynomial operators are defined in a similar way.

The operator $P_n(f)$, which assigns to each $f \in C[-1, +1]$ its unique polynomial of best approximation, *is not linear* [see Chapter 2, Sec. 9, Note 1]. It follows that there cannot exist a polynomial operator L_n of degree n with the property

$$\| f - L_n(f) \| = E_n(f), \qquad f \in C. \tag{1}$$

But if (1) is impossible, we still can hope that similar but weaker statements are true, either for the whole space C or for some of its subsets. This will be the subject of Secs. 1 and 3 of this chapter. Among other things, we shall see (Theorem 9) that $\| f - L_n(f) \| \leqslant M E_n^*(f)$, $f \in C^*$, $n = 1, 2, \cdots$, is impossible for each fixed constant M.

However, if the degree of the approximating operator is increased, the last inequality becomes feasible.

THEOREM 1 (de la Vallée-Poussin [16]). For each positive integer n, there exists a trigonometric polynomial operator $V_n(f)$ of degree $2n - 1$ with the property

$$\| f - V_n(f) \| \leqslant 4 E_n^*(f), \qquad f \in C^*. \tag{2}$$

To find the operators V_n, we compare the properties of the Fourier sums $s_n(f)$ and the Fejér means $\sigma_n(f)$ (Chapter 1, Sec. 2). The operators s_n leave invariant all trigonometric polynomials T_n of degree n. The σ_n do not have this property, but as a compensation, they satisfy $\sigma_n(f) \to f$, that is, $\| f - \sigma_n(f) \| \to 0$ for all $f \in C^*$. To obtain operators with both these properties, we consider the "gliding" average:

$$\frac{1}{p} [s_n(f, x) + s_{n+1}(f, x) + \cdots + s_{n+p-1}(f, x)]. \tag{3}$$

If $n = 0$, this is σ_p; if $p = 1$, this is s_n. But we want $p = p(n) \to \infty$. In particular, we choose $p = n$:

$$V_n(f, x) = \frac{1}{n} [s_n(f, x) + \cdots + s_{2n-1}(f, x)] = 2\sigma_{2n}(f, x) - \sigma_n(f, x). \qquad (4)$$

We note some simple properties of the operators V_n:

1. V_n is a trigonometric polynomial operator of degree $2n - 1$, and $V_n(T_n) = T_n$. This follows from the first relation (4).

2. We have $V_n(f) \to f$, for $n \to \infty$, $f \in C^*$. This follows from the second relation (4), since $\sigma_n(f) \to f$.

3. $\| V_n \| \leqslant 3$. Indeed, $\| V_n \| \leqslant 2 \| \sigma_{2n} \| + \| \sigma_n \| = 3$ (Chapter 1, Sec. 2).

We prove (2) with the help of the polynomials of best approximation T_n of the function f:

$$\| f - V_n(f) \| = \| f - T_n - V_n(f - T_n) \| \leqslant \| f - T_n \| + \| V_n \| \cdot \| f - T_n \|$$
$$\leqslant 4 \| f - T_n \| = 4E_n^*(f). \qquad \blacksquare$$

It is easy to generalize Theorem 1 somewhat. Let $0 < \epsilon < 1$ be given. Taking $p = [\epsilon n]$ in (3), we obtain a sequence of polynomial operators V_n' of degrees $\leqslant (1 + \epsilon) n$, for which $\| f - V_n'(f) \| \leqslant \text{const } E_n^*(f)$.

As an example of an application of Theorem 1, we derive

THEOREM 2. If $g(x) = | \cos x |$, then $E_n^*(g) \geqslant Mn^{-1}$, for some const $M > 0$.

Proof. The Fourier series of $g(x)$ converges. It can be easily found:

$$g(x) = \frac{2}{\pi} + \frac{4}{\pi} \sum_{k=1}^{\infty} (-1)^{k+1} \frac{\cos 2kx}{4k^2 - 1}.$$

Since

$$0 = g\left(\frac{\pi}{2}\right) = \frac{2}{\pi} - \frac{4}{\pi} \sum_{k=1}^{\infty} (4k^2 - 1)^{-1},$$

we obtain for some $M' > 0$,

$$s_m\left(g, \frac{\pi}{2}\right) = \frac{2}{\pi} - \frac{4}{\pi} \sum_{2k \leqslant m} (4k^2 - 1)^{-1} = \frac{4}{\pi} \sum_{2k > m} (4k^2 - 1)^{-1} \geqslant \frac{M'}{m}.$$

Therefore,

$$V_n\left(g, \frac{\pi}{2}\right) - g\left(\frac{\pi}{2}\right) = \frac{1}{n} \sum_{m=n}^{2n-1} s_m\left(g, \frac{\pi}{2}\right) \geqslant \frac{M'}{2n}, \qquad n = 1, 2, \cdots$$

Now let $h(x) = |x|$ on $[-1, +1]$. Then

$$E_n(h) \approx n^{-1}. \tag{5}$$

For the proof, let $P_n(x)$ be an arbitrary polynomial of degree n. If $x = \cos t$, then $|h(x) - P_n(x)| = ||\cos t| - P_n(\cos t)| \geqslant Mn^{-1}$ for at least one x, by Theorem 2. Hence, $E_n(h) \geqslant Mn^{-1}$. The opposite inequality follows from Jackson's theorem, or from Theorem 1, Chapter 5. (Actually, $nE_n(h)$ converges to a certain limit as $n \to \infty$ [2, vol. I, p. 157].)

Our next theorem concerns positive polynomial operators. For such operators, the critical degree of approximation is $O(n^{-2})$; no assumption about the smoothness of the function can assure better approximation. Let $f_0(x) = 1$, $f_1(x) = x$, $f_2(x) = x^2$.

THEOREM 3 (Korovkin). Let $L_n(f)$, $n = 1, 2, \cdots$, be positive polynomial operators of degree n, defined on $C[-1, +1]$. Then, for at least one $i = 0, 1, 2$,

$$\|f_i - L_n(f_i)\| \neq o(n^{-2}), \qquad n \to \infty. \tag{6}$$

Proof. We shall use the inequality of Cauchy,

$$|L(fg, x)| \leqslant L(f^2, x)^{1/2} L(g^2, x)^{1/2}, \tag{7}$$

which is valid[1] for positive linear operators L. Let $h(x) = |x|$, $h_a(x) = |x - a|$; then $h_a^2 = a^2 f_0 - 2af_1 + f_2$. We start with the inequality $||x| - |a|| \leqslant |x - a|$.

In terms of our functions, this inequality can be written in the form $||a| f_0 - h| \leqslant h_a$. Hence, by the properties of positive operators (Chapter 1, Sec. 2),

$$||a| L_n(f_0, x) - L_n(h, x)| \leqslant L_n(h_a, x).$$

For $x = a$, by (7),

$$||a| L_n(f_0, a) - L_n(h, a)| \leqslant L_n(h_a, a) \leqslant L_n(f_0, a)^{1/2} L_n(h_a^2, a)^{1/2}$$
$$= L_n(f_0, a)^{1/2} \{a^2[L_n(f_0, a) - f_0(a)] - 2a[L_n(f_1, a) - f_1(a)]$$
$$+ [L_n(f_2, a) - f_2(a)]\}^{1/2}. \tag{8}$$

We should like to estimate $|h(a) - L_n(h, a)|$. We have

$$|h(a) - L_n(h, a)| \leqslant |a| |f_0(a) - L_n(f_0, a)| + ||a| L_n(f_0, a) - L_n(h, a)|. \tag{9}$$

---

[1] For the proof, one considers for a fixed x the quadratic polynomial in λ:

$$L((f + \lambda g)^2, x) > 0.$$

Assume that $\| f_i - L_n(f_i) \| = o(n^{-2})$ holds for $i = 0, 1, 2$. Then, first of all, $L_n(f_0, a) \to f_0(a) = 1$, and therefore, using (8), we see that the last term in (9) is $o(n^{-1})$. This gives

$$| h(a) - L_n(h, a) | = o(n^{-2}) + o(n^{-1}) = o(n^{-1}).$$

uniformly in a for $-1 \leqslant a \leqslant 1$. But this contradicts (5). ▌

2. The Principle of Uniform Boundedness

Let L_n be a sequence of bounded, linear operators, all of which map a given Banach space X into another Banach space Y. (The X, Y do not depend on n.) What are the conditions for the convergence, or for the boundedness, of the sequence $L_n(f)$, for each $f \in X$, in the norm of the space Y? A very important *necessary* condition for the convergence is the boundedness of the norms $\| L_n \|$. This follows from the following theorem.

THEOREM 4 ("The Principle of Uniform Boundedness"). The sequence $L_n(f)$ has, for each $f \in X$, bounded norms in the space Y if and only if, for some $M \geqslant 0$,

$$\| L_n \| \leqslant M, \qquad n = 1, 2, \cdots. \tag{1}$$

Proof. It is clear that condition (1) is sufficient. We have to show that it is also necessary. If the norms are not bounded, we can assume, by passing to a subsequence, that $\| L_n \| \geqslant 10^n$, $n = 1, 2, \cdots$. We shall derive from this the existence of an $f \in X$ for which

$$\| L_n(f) \| \geqslant 8^{-1} \left(\tfrac{10}{9} \right)^n, \qquad n = 1, 2, \cdots, \tag{2}$$

and from this our theorem will follow.

According to the definition of $\| L_n \|$, we can find $f_n \in X$, $n = 1, 2, \cdots$, with

$$\| L_n(f_n) \| \geqslant \tfrac{1}{2} \| L_n \|; \qquad \| f_n \| = 1. \tag{3}$$

We choose $f = \sum\limits_{n=1}^{\infty} 9^{-n} \epsilon_n f_n$; the numbers ϵ_n, which are allowed to take only the values 0 and 1, will be selected in a moment. The series $\sum 9^{-n} \epsilon_n \| f_n \|$ converges, and this together with the completeness of the space X implies the convergence (in the norm of X) of the series defining f.

The ϵ_n are selected by induction in the following fashion: We take $\epsilon_1 = 1$. If $\epsilon_1, \cdots, \epsilon_{n-1}$ have been already selected, we take $\epsilon_n = 1$ if

$$\left\| L_n \left(\sum_{k=1}^{n-1} 9^{-k} \epsilon_k f_k \right) \right\| \leqslant \tfrac{1}{4} 9^{-n} \| L_n \| \tag{4}$$

and $\epsilon_n = 0$ if (4) does not hold. In the first case, by (3),

$$\left\| L_n \left(\sum_{k=1}^{n} 9^{-k} \epsilon_k f_k \right) \right\| \geq \| L_n(9^{-n} f_n) \| - \left\| L_n \left(\sum_{k=1}^{n-1} 9^{-k} \epsilon_k f_k \right) \right\|$$

$$\geq \tfrac{1}{2} 9^{-n} \| L_n \| - \tfrac{1}{4} 9^{-n} \| L_n \|$$

$$= \tfrac{1}{4} 9^{-n} \| L_n \|,$$

and the left-hand side is greater than $\tfrac{1}{4} 9^{-n} \| L_n \|$ also in the second case. Therefore, for each n,

$$\| L_n(f) \| \geq \left\| L_n \left(\sum_{k=1}^{n} 9^{-k} \epsilon_k f_k \right) \right\| - \left\| L_n \left(\sum_{k=n+1}^{\infty} 9^{-k} \epsilon_k f_k \right) \right\|$$

$$\geq \tfrac{1}{4} 9^{-n} \| L_n \| - \| L_n \| \sum_{k=n+1}^{\infty} 9^{-k}$$

$$= (\tfrac{1}{4} - \tfrac{1}{8}) 9^{-n} \| L_n \| \geq \tfrac{1}{8} (\tfrac{10}{9})^n. \qquad ▮$$

The partial sums $s_n(f, x)$ of the Fourier series of $f \in C^*$ for a fixed x are linear functionals (that is, linear operators that map C^* into R). Their norms tend to infinity (Chapter 1, Theorem 2). Therefore

THEOREM 5.   For each $x \in K$, there is a continuous 2π-periodic function whose Fourier series diverges at x.

(If we applied Theorem 4 to the $s_n(f)$, considered as *operators* from C^* to itself, we would obtain a weaker result, namely: There is a function $f \in C^*$ whose Fourier series is not uniformly convergent.)

3. Operators that Preserve Trigonometric Polynomials

We shall say that a trigonometric polynomial operator L_n of degree n belongs to the class $\mathfrak{T}_n$ if it preserves each trigonometric polynomial of degree $n : L_n(T_n) = T_n$. The simplest element of $\mathfrak{T}_n$ is $s_n(f)$, the partial sum of the Fourier series of f. We show that s_n is extremal in the class $\mathfrak{T}_n$ in the sense that it has the smallest norm among all $L_n \in \mathfrak{T}_n$. This leads to the result (due to Faber [48], Nikolaev, Lozinskiĭ; see Theorem 8) that for arbitrary operators $L_n \in \mathfrak{T}_n$, $n = 1, 2, \cdots$ the sequence $L_n(f)$ cannot converge on the whole space C^*. (It is still not known whether one can guarantee the existence of $x \in K$, $f \in C^*$ for which $L_n(f; x)$ is divergent.)

By f_a we denote the $a-$ *translation* of the function $f \in C^*: f_a(x) = f(x + a)$. Clearly, $\| f_a \| = \| f \|$, and $(f + g)_a = f_a + g_a$. Also, f_a is a continuous function of a: $\| f_a - f_{a_0} \| \to 0$ as $a \to a_0$, uniformly for each f.

THEOREM 6.   For each operator $L_n \in \mathfrak{T}_n$, one has the formula

$$\frac{1}{2\pi} \int_{-\pi}^{\pi} L_n(f_t , x - t) \, dt = s_n(f, x), \qquad f \in C^*, \qquad x \in K. \qquad (1)$$

This is Berman's [34] generalization of a formula of Faber and Marcinkiewicz.

Proof.   We show first that $L_n(f_t , x)$ is a continuous function of the two variables t and x. Let t_0, x_0 be fixed. Then

$$| L_n(f_t , x) - L_n(f_{t_0} , x_0) | \leqslant | L_n(f_{t_0} , x) - L_n(f_{t_0} , x_0) |$$
$$+ | L_n(f_t , x) - L_n(f_{t_0} , x) | .$$

The first term on the right is small if x is close to x_0 because $L_n(f_{t_0} , x)$ is a continuous function of x. The second term is small if t is close to t_0 because it does not exceed $\| L_n \| \cdot \| f_t - f_{t_0} \|$.

We see that the integrand in (1) is a continuous function of t, and thus the integral (1) exists. We denote by $A_n(f, x)$ the left-hand side of (1), and prove the identity $A_n(f) = s_n(f)$ for the following classes of functions:

(a)   Let $f = T_n$. Then $f_t = (T_n)_t = T_{nt}$ is also a trigonometric polynomial of degree n; hence, $L_n(T_{nt}) = T_{nt}$ and $L_n(T_{nt} , x - t) = T_{nt}(x - t) = T_n(x)$. Since also $s_n(T_n , x) = T_n(x)$, (1) follows.

(b)   This relation holds for $g(x) = \cos px$, $h(x) = \sin px$ if p is an integer and $p > n$. For example, consider the function g. Since

$$g_t(x) = g(x) \cos pt - h(x) \sin pt,$$

$$A_n(g, x) = \frac{1}{2\pi} \int_{-\pi}^{\pi} \{ L_n(g, x - t) \cos pt - L_n(h, x - t) \sin pt \} \, dt = 0$$

because $\cos pt$ and $\sin pt$ are orthogonal to $L_n(g, x - t)$ and $L_n(h, x - t)$, which are trigonometric polynomials of degree n in t. Thus, (1) holds, both sides being zero.

(c)   From (a) and (b) it follows that $A_n(f) = s_n(f)$ is true for all trigonometric polynomials f, whatever their degree. These form a dense set in C^*. To complete the proof, it is sufficient to show that both $A_n(f)$ and $s_n(f)$ are continuous in the norm of C^*. Now $s_n(f)$ is linear and bounded (Chapter 1, Sec. 2) and hence continuous; $A_n(f)$ is also linear and

$$| A_n(f, x) | \leqslant \frac{1}{2\pi} \int_{-\pi}^{\pi} | L_n(f_t , x - t) | \, dt$$

$$\leqslant \frac{1}{2\pi} \int_{-\pi}^{\pi} \| L_n \| \cdot \| f_t \| \cdot dt = \| L_n \| \cdot \| f \| , \qquad x \in K,$$

so that

$$\| A_n \| \leqslant \| L_n \| . \qquad \blacksquare \qquad (2)$$

' From Theorem 6 and (2) we can derive several important corollaries.

THEOREM 7.   If $L_n \in \mathfrak{T}_n$, then $\| L_n \| \geqslant \| s_n \|$.

THEOREM 8.   If $L_n \in \mathfrak{T}_n$, $n = 1, 2, \cdots$, then there exists an $f \in C^*$ for which $L_n(f)$ does not converge (and is not bounded).

This follows from the fact that $\| s_n \| \geqslant A \log n$ (Chapter 1, Theorem 2) and the Theorems 7 and 4.   ▌

THEOREM 9.   Let L_n be a sequence of trigonometric polynomial operators of degree n. There cannot exist a continuous function $\phi(u) \geqslant 0$ with $\phi(0) = 0$, for which

$$\| f - L_n(f) \| \leqslant \phi(E_n^*(f)), \qquad f \in C^*, \qquad n = 1, 2, \cdots. \tag{3}$$

Proof.   It follows from (3) that $L_n(T_n) = T_n$, so that $L_n \in \mathfrak{T}_n$. Again from (3), $L_n(f) \to f$ as $n \to \infty$ for $f \in C^*$, and this contradicts Theorem 8.   ▌

4. Trigonometric Saturation Classes

A sequence of linear operators L_n that maps C^* into itself is called *saturated* if there exists an "optimal" order $\psi(n)$ of approximation of functions $f \in C^*$ by $L_n(f)$ such that better approximation occurs only for very special functions; for example, for constants. Only the last case will be essential in this section. To give a formal definition, we require the existence of a function $\psi(n) > 0$, $n = 1, 2, \cdots$, which tends to zero as $n \to \infty$, such that

$$\lim_{n \to \infty} \frac{\| f - L_n(f) \|}{\psi(n)} = 0 \tag{1}$$

implies that f is a constant; and there should exist at least one nonconstant function $f \in C^*$ with $\| f - L_n(f) \| = O(\psi(n))$. The function ψ is then called an *optimal degree of approximation*, and the set S of functions f that satisfies the last relation is the *saturation class* of the L_n relative to ψ.

For example, *the Fejér sums $\sigma_n(f)$ [1.2(12)] are saturated with $\psi(n) = n^{-1}$.* Assume that $\| f - \sigma_n(f) \| = o(n^{-1})$ for some subsequence of the n. For the Fourier series of f,

$$\frac{a_0}{2} + \sum_{k=1}^{\infty} (a_k \cos kx + b_k \sin kx) \equiv \sum_0^{\infty} A_k(x), \tag{2}$$

the arithmetic means σ_n of the partial sums are

$$\sigma_n(x) = \sum_0^n \frac{n - k}{n} A_k(x).$$

Hence,

$$\frac{1}{\pi} \int_{-\pi}^{\pi} \sigma_n(x) \begin{Bmatrix} \cos kx \\ \sin kx \end{Bmatrix} dx = \frac{n-k}{n} \begin{Bmatrix} a_k \\ b_k \end{Bmatrix}, \qquad k \leqslant n. \tag{3}$$

From this and 1.2(8) we have

$$\frac{k}{n} a_k = \pi^{-1} \int_{-\pi}^{\pi} (f - \sigma_n) \cos kx \, dx, \qquad \sigma_n = \sigma_n(f).$$

Hence $(k/n) a_k = o(n^{-1})$ for our subsequence of n, so that $a_k = 0$ and similarly $b_k = 0$ for $k = 1, 2, \cdots$. Thus, f reduces to the constant $\frac{1}{2} a_0$. But for the function $f(x) = \cos x$, one has $\sigma_n = (1 - n^{-1}) \cos x$ and $\| f - \sigma_n(f) \| = n^{-1}$. This proves our statement. On the other hand, it is easy to see that the Fourier sums $s_n(f)$ are *not saturated* (see Problem 2).

To determine the saturation class of the $\sigma_n(f)$, we need the notion of the conjugate function. Let (2) be the Fourier series of an integrable function f. Then

$$\sum_{k=1}^{\infty} (b_k \cos kx - a_k \sin kx) \equiv \sum_{k=1}^{\infty} B_k(x) \tag{4}$$

is called the *conjugate series of f*. Let $\tilde{s}_n(f, x)$, $n = 1, 2, \cdots$, denote the partial sums of (4), $\tilde{\sigma}_n(f, x) = (\tilde{s}_1 + \cdots + \tilde{s}_{n-1})/n$, $n = 2, 3, \cdots$, their arithmetic means. We wish to express $\tilde{s}_n$, $\tilde{\sigma}_n$ explicitly in terms of f.

The following formulas, similar to 1.2(10) and 1.2(11), are useful:

$$\tilde{D}_n(x) = \sin x + \cdots + \sin nx = \frac{\cos \frac{1}{2} x - \cos (n + \frac{1}{2}) x}{2 \sin \frac{1}{2} x},$$

$$\tilde{F}_n(x) = \frac{\tilde{D}_1(x) + \cdots + \tilde{D}_{n-1}(x)}{n} = \frac{1}{2} \cot \frac{1}{2} x - \frac{\sin nx}{4n \sin^2 \frac{1}{2} x}. \tag{5}$$

We leave their proof to the reader. Since

$$B_k(x) = \pi^{-1} \int_{-\pi}^{\pi} f(x + t) \sin kt \, dt,$$

we obtain

$$\tilde{s}_n(f, x) = \frac{1}{\pi} \int_0^{\pi} [f(x + t) - f(x - t)] \tilde{D}_n(t) \, dt,$$

$$\tilde{\sigma}_n(f, x) = \frac{1}{\pi} \int_0^{\pi} [f(x + t) - f(x - t)] \tilde{F}_n(t) \, dt. \tag{6}$$

We shall say that

$$\tilde{f}(x) = \frac{1}{2\pi} \int_0^{\pi} [f(x + t) - f(x - t)] \cot \frac{1}{2} t \, dt \tag{7}$$

is the *conjugate function* of $f \in C^*$, if the integral (7) converges absolutely for all

$x \in K$ and if $\int_0^\pi |f(x + t) - f(x - t)| \cot \frac{1}{2} t \, dt$ is an integrable function. This is not the usual definition of the conjugate function [see 6, Note 5], but will allow us quickly to characterize the saturation class.

LEMMA 1.   (a) If $\tilde{f}$ exists, then the series (4) is its Fourier series. (b) If $f \in$ Lip α for some $\alpha > 0$, then $\tilde{f}$ exists and is the sum of the series (4).

Proof.   Multiplying (7) by cos kx, integrating, and changing the order of integrations, we have

$$
\tilde{a}_k = \frac{1}{\pi} \int_{-\pi}^{\pi} \tilde{f}(x) \cos kx \, dx
$$

$$
= \frac{1}{2\pi^2} \int_0^\pi \cot \frac{1}{2} t \, dt \int_{-\pi}^{\pi} f(u) \left[\cos k(u - t) - \cos k(u + t) \right] du
$$

$$
= b_k \frac{1}{\pi} \int_0^\pi \frac{\sin kt \cos \frac{1}{2} t \, dt}{\sin \frac{1}{2} t}
$$

$$
= b_k \frac{1}{\pi} \int_0^\pi (D_k + D_{k-1}) \, dt = b_k, \qquad k = 1, 2, \cdots,
$$

and $\tilde{a}_0 = 0$; similarly $\tilde{b}_k = - a_k$, $k = 1, 2, \cdots$.

Now let $f \in$ Lip α. The existence of $\tilde{f}$ is obvious. The difference between $\tilde{f}(x)$ and $\tilde{s}_n(f, x)$ is

$$
I_n = \frac{1}{2\pi} \int_0^\pi g(t) \cos (n + \tfrac{1}{2}) t \, dt,
$$

where $g(t)$ is the absolutely integrable function $[f(x + t) - f(x - t)]/\sin \frac{1}{2} t$. By Riemann-Lebesgue theorem (see [24], p. 24), we have $I_n \to 0$ for each x. ∎

After these preparations, we can formulate the following result.

THEOREM 10 (M. Zamansky).   The saturation class S of the operators σ_n consists of all functions $f \in C^*$ that have a conjugate function $\tilde{f} \in$ Lip 1.

Proof.   (a)   Let $\| f - \sigma_n(f) \| \leqslant Mn^{-1}$. The function $\varDelta_n = f - \sigma_n(f)$ has the Fourier series $\sum_{k=0}^{n} (k/n) A_k + \sum_{k=n+1}^{\infty} A_k$. We apply to $\varDelta_n$ the operator σ_N, with $N \leqslant n$:

$$
\sigma_N(\varDelta_n) = \sum_{k=1}^{N} \left(1 - \frac{k}{N} \right) \frac{k}{n} A_k .
$$

Since $\| \sigma_N \| = 1$ (Chapter 1, Sec. 2), we see that $\| \sigma_N(\varDelta_n) \| \leqslant Mn^{-1}$; hence

$$
\left\| \sum_{k=1}^{N} \left(1 - \frac{k}{N} \right) k A_k \right\| \leqslant M, \qquad N = 1, 2, \cdots, n. \tag{8}
$$

But $- kA_k = B'_k$ is the derivative of the general term of the series (4); therefore, (8) is equivalent to $\| \tilde{\sigma}'_n(f) \| \leqslant M$, $n = 2, 3, \cdots$. For the $\tilde{\sigma}_n$ themselves, this implies $| \tilde{\sigma}_n(x + t) - \tilde{\sigma}_n(x) | \leqslant M | t |$, $n = 2, 3, \cdots$. According to Theorem 6, Chapter 4, $f \in \mathrm{Lip}\ \alpha$ if $0 < \alpha < 1$, and therefore $\tilde{f}$ exists. By Lemma 1, $\tilde{\sigma}_n(x) \to \tilde{f}(x)$ for each $x \in K$. Hence, $| \tilde{f}(x + t) - \tilde{f}(x) | \leqslant M | t |$.

(b) Conversely, let $| \tilde{f}(x + t) - \tilde{f}(x) | \leqslant M | t |$ for some M. We must show that $f \in S$. According to Lemma 1, $-f + \frac{1}{2} a_0$ and $\tilde{\tilde{f}}$ have the same Fourier series. The arithmetic means of this series converge to $-f + \frac{1}{2} a_0$ (p. 10) and to $\tilde{\tilde{f}}$ [by Lemma 1, (b)]; hence, the two functions are identical. Interchanging f and $\tilde{f}$ in (6) and (7), we have

$$f(x) - \sigma_n(f, x) = - \frac{1}{\pi} \int_0^\pi [\tilde{f}(x + t) - \tilde{f}(x - t)] \frac{\sin nt}{4n \sin^2 (t/2)}\ dt$$

$$= I_1 + I_2 , \tag{9}$$

say, where the integrals I_1, I_2 are over the intervals $(0, \pi/n)$ and $(\pi/n, \pi)$, respectively. Since $| \sin nt | \leqslant nt$,

$$| I_1 | \leqslant \frac{M}{2\pi} \int_0^{\pi/n} \frac{t^2}{\sin^2 (t/2)}\ dt = O(n^{-1}).$$

For I_2, we integrate by parts with the help of the function

$$\psi_n(t) = \int_t^\pi \frac{\sin nu}{\sin^2 (u/2)}\ du;$$

$$| I_2 | \leqslant \frac{1}{4\pi n} \left| [\tilde{f}(x + t) - \tilde{f}(x - t)]\ \psi_n(t) \Big|_{t\ =\ \pi/n}^{t\ =\ \pi} \right|$$

$$+ \frac{1}{4\pi n} \left| \int_{\pi/n}^\pi \psi_n(t)\ d_t\{\tilde{f}(x + t) - \tilde{f}(x - t)\} \right|$$

$$\leqslant \frac{M}{2n^2} \left| \psi_n \left(\frac{\pi}{n} \right) \right| + \frac{M}{2\pi n} \int_{\pi/n}^\pi | \psi_n(t) |\ dt.$$

(The last estimation follows from the definition of the Stieltjes integral.) By the second mean-value theorem,

$$\int_a^b fg\ dt = g(a) \int_a^\xi f\ dt, \qquad a \leqslant \xi \leqslant b,$$

if g is decreasing and positive and f is continuous. Therefore,

$$| \psi_n(t) | \leqslant \frac{2}{n} \frac{1}{\sin^2 (t/2)} \leqslant \frac{2\pi^2}{nt^2} .$$

We obtain

$$| I_2 | \leqslant \frac{M}{n} + \frac{M\pi}{n^2} \int_{\pi/n}^{\pi} \frac{dt}{t^2} \leqslant 2 \frac{M}{n} = O\left(\frac{1}{n}\right).$$

Hence,

$$\| f - \sigma_n(f) \| = O(n^{-1}). \qquad\blacksquare$$

5. The Saturation Class of the Bernstein Polynomials

As another example of a saturation class, we consider the approximation of functions $f \in C[0, 1]$ by Bernstein polynomials $B_n(f)$ [see 1.2(5)]. The results of Chapter 5 suggest that the optimal order of approximation of $f(x)$ by $B_n(f, x)$ could depend on x. That this is really so will be shown by Theorem 11. Several changes become necessary in the definitions of Sec. 4. The optimal degree of approximation $\psi(x, n) = x(1 - x)/n$ is now a function of n and x. The exceptional functions f for which $f(x) - B_n(f, x) = o(\psi(x, n))$ are now all *linear* functions.

THEOREM 11. The saturation class S of the Bernstein polynomials consists of all $f \in C[0, 1]$ for which $f' \in \text{Lip } 1$. The optimal order of approximation is $x(1 - x)/n$; more precisely, $f' \in \text{Lip}_M 1$ is equivalent to

$$| f(x) - B_n(f, x) | \leqslant M \frac{x(1 - x)}{2n}, \qquad n = 1, 2, \cdots; \qquad 0 \leqslant x \leqslant 1. \qquad (1)$$

If, in addition to (1), $| f(x) - B_n(f, x) | \leqslant n^{-1} \epsilon_n x(1 - x)$, $0 \leqslant x \leqslant 1$, where $\epsilon_n > 0$, $\underline{\lim} \epsilon_n = 0$, then f is linear.

One should compare (1) with the approximation of functions $f \in S$ available by arbitrary polynomials $P_n(x)$ of degree n. Theorem 2, Chapter 5, gives

$$| f(x) - P_n(x) | \leqslant C \max \left(\frac{x(1 - x)}{n^2}, \frac{1}{n^4} \right).$$

We see that Bernstein polynomials converge fairly slowly.

Proof. First let $f' \in \text{Lip}_M 1$. For any two points x, y of $[0, 1]$, we have

$$f(x) - f(y) = \int_y^x f'(t) \, dt = f'(x) (x - y) - \int_y^x (t - y) \, df'(t).$$

Since $| f'(t + h) - f'(t) | \leqslant M|h|$, the absolute value of the last integral does not exceed $M \int_y^x (t - y) \, dt = \frac{1}{2} M(x - y)^2$. Applying this with $y = k/n$ to the formula

$$f(x) - B_n(f, x) = \sum_{k=0}^{n} \left\{ f(x) - f\left(\frac{k}{n}\right) \right\} p_{nk}(x),$$

and taking into account 1.3(12), we obtain

$$|f(x) - B_n(f, x)| \leqslant \tfrac{1}{2} M \sum_{k=0}^{n} \left(x - \frac{k}{n}\right)^2 p_{nk}(x) = M \frac{x(1-x)}{2n}.$$

The main difficulty is to prove that (1) implies $f' \in \text{Lip}_M 1$. Let $G(x)$, $0 \leqslant x \leqslant 1$ be an arbitrary function of the following type: $G(x)$ is twice continuously differentiable on $[0, 1]$; it vanishes outside some interval (a, b), $0 < a < b < 1$. We write $g(x) = [\tfrac{1}{2} x(1-x)]^{-1} G(x)$, $0 < x < 1$, $g(0) = g(1) = 0$. With each G, we associate a sequence of linear functionals $L_n(f)$, defined for all $f \in C[0, 1]$:

$$L_n(f) = \sum_{k=0}^{n} \left[B_n\left(f, \frac{k}{n}\right) - f\left(\frac{k}{n}\right)\right] g\left(\frac{k}{n}\right). \qquad (2)$$

Our main intermediate result is

LEMMA 2.   For each $f \in C[0, 1]$ and each G, we have

$$L_n(f) \to \int_0^1 f(x)\, G''(x)\, dx. \qquad (3)$$

Let $0 < a_1 < a$, $b < b_1 < 1$. We may restrict the summation in (2) to the set I of the values of k for which $na_1 < k < nb_1$, since for $k \notin I$, $g(k/n) = 0$. Substituting the formula 1.2(5) for $B_n(f, x)$ into (2) and rearranging, we obtain

$$L_n(f) = \sum_{k \in I} \sum_{l=0}^{n} f\left(\frac{l}{n}\right) g\left(\frac{k}{n}\right) p_{nl}\left(\frac{k}{n}\right) - \sum_{l=0}^{n} f\left(\frac{l}{n}\right) g\left(\frac{l}{n}\right). \qquad (4)$$

Since g'' is continuous,

$$g(x) = g(x_0) + (x - x_0) \cdot g'(x_0) + \tfrac{1}{2}(x - x_0)^2 g''(x_0) + R(x, x_0),$$

where $|R(x, x_0)| = |x - x_0|^2 \epsilon(x, x_0)$, and $\epsilon(u, v)$ is some function that is bounded and tends to zero as $u - v \to 0$. If we take $x = k/n$, $x_0 = l/n$, we obtain from (4),

$$L_n(f) = \sum_{k \in I} \sum_{l=0}^{n} f\left(\frac{l}{n}\right) \left\{ g\left(\frac{l}{n}\right) + \left(\frac{k}{n} - \frac{l}{n}\right) g'\left(\frac{l}{n}\right) \right.$$

$$\left. + \frac{1}{2}\left(\frac{k}{n} - \frac{l}{n}\right)^2 g''\left(\frac{l}{n}\right) \right\} p_{nl}\left(\frac{k}{n}\right) - \sum_{l=0}^{n} f\left(\frac{l}{n}\right) g\left(\frac{l}{n}\right) + R_n, \qquad (5)$$

where

$$|R_n| \leqslant \text{const} \sum_{k,l=0}^{n} \left(\frac{k}{n} - \frac{l}{n}\right)^2 \epsilon\left(\frac{k}{n}, \frac{l}{n}\right) p_{nl}\left(\frac{k}{n}\right). \qquad (6)$$

We first show that $R_n \to 0$. Taking an arbitrary $\epsilon > 0$, we have $\epsilon(u, v) \leqslant \epsilon$ for all u, v with $|u - v| \leqslant \delta$, $\delta > 0$. We split the sum (6) into parts $\sum_1$ and $\sum_2$ corresponding to the inequalities

$$\left| \frac{k}{n} - \frac{l}{n} \right| < \delta \quad \text{and} \quad \left| \frac{k}{n} - \frac{l}{n} \right| \geqslant \delta.$$

Then (see Chapter 1, Sec. 3)

$$\sum_1 \leqslant \text{const } \epsilon \sum_{k=0}^{n} \frac{1}{n^2} T_{n2} \left(\frac{k}{n} \right) \leqslant \text{const } \epsilon,$$

since $T_{n2}(x) \leqslant n$ for $0 \leqslant x \leqslant 1$. For the sum $\sum_2$ we can use 1.3(13): $\sum_2 \leqslant \text{const } (n + 1) \, n^{-2} \to 0$. Thus $R_n \to 0$ as $n \to \infty$.

To investigate the first sum in (5)—which we denote by $S_n(f)$—we use the Euler-Maclaurin formula. This formula serves to estimate the difference between a Riemann sum and the integral of a function that has small higher derivatives. We can write $S_n(f) = \sum_{k \in I} Q(k/n)$, where

$$Q(x) = \sum_{l=0}^{n} f\left(\frac{l}{n}\right) \left\{ g\left(\frac{l}{n}\right) + \left(x - \frac{l}{n}\right) g'\left(\frac{l}{n}\right) + \frac{1}{2}\left(x - \frac{l}{n}\right)^2 g''\left(\frac{l}{n}\right) \right\} p_{nl}(x).$$

$$(7)$$

The Euler-Maclaurin formula for a four-times continuously differentiable function Q has the form

$$\sum_{k=m}^{p-1} Q\left(\frac{k}{n}\right) = n \int_{m/n}^{p/n} Q(t) \, dt - \frac{1}{2}\left[Q\left(\frac{p}{n}\right) - Q\left(\frac{m}{n}\right) \right]$$

$$+ \frac{1}{12n}\left[Q'\left(\frac{p}{n}\right) - Q'\left(\frac{m}{n}\right) \right]$$

$$- \frac{1}{720n^4} \sum_{k=m}^{p-1} Q^{(4)}\left(\frac{k + \theta}{n}\right), \quad 0 \leqslant \theta \leqslant 1, \quad (8)$$

(see [29, p. 128]).

Let m be the smallest and $p - 1$ the largest k in I; then (8) is equal to $S_n(f)$.

Now the function $Q(x)$ is small if $0 \leqslant x \leqslant m/n$ or $p/n \leqslant x \leqslant 1$. Indeed, if $\delta > 0$ is so chosen that $\delta < a - a_1$, $\delta < b_1 - b$, then for all large n and all x in the preceding intervals, $|x - y| \geqslant \delta$ if $a \leqslant y \leqslant b$. Since the function $g(y)$ is zero outside $[a, b]$, we derive from (7),

$$|Q(x)| = O(1) \sum_{a \leqslant l/n \leqslant b} p_{nl}(x) = O(1) \sum_{|x - l/n| \geqslant \delta} p_{nl}(x) = O(n^{-2})$$

[see 1.3(13)]. This shows that the first two terms on the right in (8) are $= n \int_0^1 Q \, dt + o(1)$.

Next, the derivatives $Q', Q^{(4)}$ in (8) are small compared with the denominators n, n^4. This follows from the following facts: For some constant C, depending on $\delta > 0$,

$$| Q'(x) | \leqslant C \sqrt{n}, \qquad | Q^{(r)}(x) | \leqslant C n^2, \qquad \delta \leqslant x \leqslant 1 - \delta, \qquad r \leqslant 4. \qquad (9)$$

For Q' we can derive this from 1.3(11): If $\delta \leqslant x \leqslant 1 - \delta$, then

$$| p'_{nl}(x) | \leqslant \text{const} \, | l - nx | \, p_{nl}(x),$$

and therefore by Cauchy's inequality,

$$\sum_{l=0}^{n} | p'_{nl}(x) | = O(1) \sum_{l=0}^{n} | l - nx | \, p_{nl}^{1/2} \cdot p_{nl}^{1/2}$$

$$= O(1) \, T_{n2}(x)^{1/2} = O(\sqrt{n}).$$

Thus,

$$| Q'(x) | = O(1) \sum_{l=0}^{n} [| p'_{nl}(x) | + p_{nl}(x)] = O(\sqrt{n}), \qquad \delta \leqslant x \leqslant 1 - \delta.$$

For larger r, we have the formula (easily proved by induction)

$$p_{nl}^{(r)}(x) = \sum_{\substack{i, j \\ i + 2j \leqslant r}} (l - nx)^i \, n^j X_{ij}(x) \, p_{nl}(x), \qquad (10)$$

where the functions $X_{ij}(x)$ have derivatives of all orders on $[\delta, 1 - \delta]$. Let $r \leqslant 4$. We see that $| p_{nl}^{(r)}(x) |$ does not exceed a sum of constant multiples of terms $| l - nx |^i n^j p_{nl}(x)$, $i + 2j \leqslant 4$. We may restrict i to be even; for if i is odd, then $i + 1 + 2j \leqslant 4$, and $| l - nx |^i n^j \leqslant | l - nx |^{i-1} n^{j+1}$. Hence, by Chapter 1, Sec. 3,

$$\sum_{l=0}^{n} | p_{nl}^{(r)}(x) | = O(1) \sum_{l=0}^{n} \{ n^2 + n(l - nx)^2 + (l - nx)^4 \} \, p_{nl}(x)$$

$$= O(1) \{ n^2 + n T_{n2}(x) + T_{n4}(x) \} = O(n^2),$$

for $\delta \leqslant x \leqslant 1 - \delta$. This allows us to deduce the second inequality (9).

If $\delta < a_1, b_1 < 1 - \delta$, then $p/n, m/n$ are in $(\delta, 1 - \delta)$ for all large n, and we obtain from (8),

$$S_n(f) = n \int_0^1 Q(x) \, dx + o(1). \qquad (11)$$

We compute this integral. With the help of the beta function, we have

$$\int_0^1 p_{nl}(x) \, dx = \binom{n}{l} B(l + 1, n - l + 1) = \frac{1}{n + 1},$$

and similarly

$$\int_0^1 x p_{nl}(x)\,dx = \frac{l+1}{(n+1)(n+2)},$$

$$\int_0^1 x^2 p_{nl}(x)\,dx = \frac{(l+1)(l+2)}{(n+1)(n+2)(n+3)}.$$

Therefore

$$\int_0^1 \left(x - \frac{l}{n}\right) p_{nl}(x)\,dx = \frac{n-2l}{n(n+1)(n+2)} = \frac{1}{n(n+1)}\left(1 - 2\frac{l}{n}\right) + O(n^{-3}),$$

$$\int_0^1 \left(x - \frac{l}{n}\right)^2 p_{nl}(x)\,dx = \frac{1}{n(n+1)}\left(\frac{l}{n} - \frac{l^2}{n^2}\right) + O(n^{-3}).$$

Using (5), (11), and (7), we obtain

$$L_n(f) = S_n(f) - \sum_{l=0}^n f\left(\frac{l}{n}\right) g\left(\frac{l}{n}\right) + o(1)$$

$$= \frac{1}{n+1}\sum_{l=0}^n f\left(\frac{l}{n}\right)$$

$$\times \left\{-g\left(\frac{l}{n}\right) + \left(1 - 2\frac{l}{n}\right) g'\left(\frac{l}{n}\right) + \tfrac{1}{2}\left(\frac{l}{n} - \frac{l^2}{n^2}\right) g''\left(\frac{l}{n}\right)\right\} + o(1)$$

$$= \frac{1}{n+1}\sum_{l=0}^n f\left(\frac{l}{n}\right) G''\left(\frac{l}{n}\right) + o(1) \to \int_0^1 f(x)\, G''(x)\,dx. \qquad \blacksquare$$

After Lemma 2 has been established, the proof of Theorem 11 is easily completed. For functions f which satisfy (1), we obtain another formula for the limit of a sequence of the $L_n(f)$. From the definition of the function $g(x)$, it follows that $L_n(f)$ is equal to the Stieltjes integral,

$$L_n(f) = \int_0^1 G(x)\, d\lambda_n(x), \qquad (12)$$

where the function $\lambda_n(x)$, defined on $[0, 1]$, has jumps equal to

$$\sigma_k = 2\left[\frac{k}{n}\left(1 - \frac{k}{n}\right)\right]^{-1}\left\{B_n\left(f, \frac{k}{n}\right) - f\left(\frac{k}{n}\right)\right\}$$

at the points $x = k/n$, $k = 1, \cdots, n - 1$, and is constant between these points. We can take

$$\lambda_n(x) = \sum_{\substack{1 \leqslant k \leqslant n-1 \\ k < nx}} \sigma_k.$$

The total variation of $\lambda_n(x)$ is bounded by M because $|\sigma_k| \leqslant Mn^{-1}$, according to (1). Since $\lambda_n(0) = 0$, the functions $\lambda_n(x)$ are uniformly bounded. By a theorem of Helly [26, p. 222], one can extract from $\lambda_n(x)$ a subsequence $\lambda_{n_p}(x)$ that converges everywhere on [0, 1] to a function $\lambda(x)$.

For two points x, y in [0, 1], the absolute value of the difference $\lambda_n(x) - \lambda_n(y)$ does not exceed Mn^{-1} times the number of points k/n between x and y. This number is at most $n|x - y| + 1$; hence, $|\lambda_n(x) - \lambda_n(y)| \leqslant M|x - y| + Mn^{-1}$. It follows that $|\lambda(x) - \lambda(y)| \leqslant M|x - y|$ or $\lambda \in \mathrm{Lip}_M 1$.

By another theorem of Helly [26, p. 233],

$$\int_0^1 G(x)\, d\lambda_{n_p}(x) \to \int_0^1 G(x)\, d\lambda(x). \tag{13}$$

The limits in (3) and (13) must be equal; thus, by partial integration applied twice,

$$\int_0^1 f G'' dx = \int_0^1 \Lambda G'' dx, \tag{14}$$

where Λ is an integral of λ. We use the following

LEMMA 3.   If $f \in C[a, b]$, and if

$$\int_a^b f G'' dx = 0 \tag{15}$$

holds for each twice continuously differentiable G on $[a, b]$ such that G, G', G'' vanish at a, b, then the function f is linear.

It is sufficient to show that for each $h > 0$, the difference $\Delta_h f(x)$ does not depend on x. Otherwise, there are distinct points x_i, $i = 1, \cdots, 4$ inside $[a, b]$ for which $x_4 - x_3 = x_2 - x_1 > 0$ and $f(x_1) - f(x_2) - f(x_3) + f(x_4)$ is not zero; say, is > 0. We take $\delta > 0$ so small that the intervals $(x_i - \delta, x_i + \delta)$ are disjoint and contained in $[a, b]$. We put $h(x_1) = h(x_4) = 1$, $h(x_2) = h(x_3) = -1$, $h(a) = h(b) = h(x_i \pm \delta) = 0$, and let h be linear between all these points. It is easy to see that h has a second integral that satisfies the requirements for G. Moreover,

$$\delta^{-1} \int_{x_i-\delta}^{x_i+\delta} f h\, dx \to 2 f(x_i)\, h(x_i) \qquad \text{as} \qquad \delta \to 0,$$

so that $\int_a^b f h\, dx > 0$ for small $\delta > 0$, a contradiction. |

From (14) and Lemma 3 we derive that $f = \Lambda + Ax + B$; hence, $f' = \lambda + A \in \mathrm{Lip}_M 1$ on $[a, b]$. Since a, b, $0 < a < b < 1$ are arbitrary, $f' \in \mathrm{Lip}_M 1$ on [0, 1].

The last part of Theorem 11 follows easily. We have

$$|f(x) - B_n(f, x)| \leqslant \epsilon \frac{x(1-x)}{2n}, \qquad \text{for each} \qquad \epsilon > 0$$

and infinitely many n. From this we can again deduce that $f' \in \mathrm{Lip}_\epsilon 1$. Since $\epsilon > 0$ is arbitrary, we have $f' = \mathrm{const}$.   ❙

6. Notes

1. Condition 2(1) of Theorem 4 is only necessary for the convergence of the sequence $L_n(f)$ for each $f \in X$. But it is very close to being a necessary and sufficient condition. More exactly, a sequence of bounded linear operators $L_n(f)$ that map a Banach space X into another Banach space Y converges in the norm for each $f \in X$ if and only if: (a) it converges for some set of elements of X, whose linear combinations are dense in X; (b) the norms $\| L_n \|$ are bounded. This theorem is a standard source of special results about convergent sequences of operators.

2. For functions $f \in C^*$, it is natural to consider sequences of polynomial operators:

$$L_n(f, x) = \lambda_0^{(n)} \frac{a_0}{2} + \sum_{k=1}^{n} \lambda_k^{(n)} (a_k \cos kx + b_k \sin kx),$$

where a_k, b_k are the Fourier coefficients of f. The constants $\lambda_k^{(n)}$ determine the sequence L_n. Under what conditions, in terms of the $\lambda_k^{(n)}$, is it true that $L_n(f) \to f$ for each $f \in C^*$? By means of the theorem stated in note 1, we get the following necessary and sufficient conditions [15, p. 485]: (a) $\lambda_k^{(n)} \to 1$ as $n \to \infty$ for each fixed k; (b) the "Lebesgue constants"

$$A_n = \int_0^\pi \left| \tfrac{1}{2} \lambda_0^{(n)} + \sum_{k=1}^{n} \lambda_k^{(n)} \cos kt \right| dt$$

are bounded.

3. For special $\lambda_k^{(n)}$, one can give simpler conditions for the $\lambda_k^{(n)}$. Here, the most interesting result is due to Nikol'skiĭ. He showed [15, p. 504] that if the second differences $\lambda_k^{(n)} - 2\lambda_{k+1}^{(n)} + \lambda_{k+2}^{(n)}$ (we put $\lambda_l^{(n)} = 0$ if $l > n$) are all positive or all negative, then the following conditions are necessary and sufficient: condition (a) of Note 2;

(b) $| \lambda_k^{(n)} | \leqslant M$ for all k, n;

and

(c) $\displaystyle\sum_{k=0}^{n} \lambda_k^{(n)} (n - k + 1)^{-1} = O(1).$

4. The simple proof of Theorem 10 given by Favard [50] depends almost entirely upon the general theory of series, and reduces to a minimum the use of specific properties of Fourier series. Our proof, however, has the advantage

of being applicable to other sequences of operators. General methods for the determination of saturation classes have been given by Butzer, Sunouchi and Watari [92], and Tureckiĭ [99]. De Leeuw was the first to discuss saturation classes of Bernstein polynomials. His results do not yet take into account the factor $x(1 - x)$. Theorem 11 is given in Lorentz [76].

5.   The definition of $\tilde{f}$, adopted in Sec. 4, is not the standard one. Usually, $\tilde{f}(x)$ is said to be defined if the integral (7) converges (for all $x \in K$) in the Cauchy sense, that is, if the limit of $\int_h^\pi$ as $h \to 0 +$ exists. With this definition of $\tilde{f}$, Theorem 10 is also true, but requires deeper lemmas about conjugate functions.

PROBLEMS

1.   Show that there does not exist a sequence of *algebraic* polynomial operators $L_n(f)$ with the property $L_n(f) \to f$ for each $f \in C[a, b]$.

2.   The operator $s_n(f)$ has no optimal order of approximation.

3.   For $f \in C[a, b]$, let $z(f)$ denote the (finite or infinite) number of zeros of $f(x)$ in $[a, b]$. A linear operator $g = L(f)$ has the property (Z) if $z(g) \leqslant z(f)$ for each $f \in C$. If, in addition, L transforms the function 1 into itself, then

$$\operatorname*{Var}_{a,b} L(f) \leqslant \operatorname*{Var}_{a,b} f \text{ for each } f \in C.$$

(*Hint:* Use Banach's formula for the variation, [26, p. 225].)

4.   The Bernstein operator $B_n(f)$ [1.2(5)] has the property (Z) for each n. (Schoenberg) [*Hint*: Use Descartes' rule of signs.]

5.   Theorem 3 remains true for positive *trigonometric* operators if the functions 1, x, x^2 are replaced by 1, $\sin x$, $\cos x$ (Korovkin).

6.   If L_n, $n = 1, 2, \cdots$ are positive linear functionals on $C[a, b]$, and if $L_n(e) \to 1$, $L_n(g) \to 0$ for the functions $e(x) = 1$, $g(x) = (x - c)^2$, $a \leqslant c \leqslant b$, then $L_n(f) \to f(c)$ for all $f \in C[a, b]$ (Korovkin).

7.   Prove the theorem formulated in Sec. 6, Note 1. As an application, derive Theorem 5, Chapter 1, from the properties of the Fejér kernel.

8.   Prove that

$$V_{m,n}(f, x) = \frac{1}{n - m} \{s_m(f, x) + \cdots + s_{n-1}(f, x)\}, \qquad m < n,$$

is equal to the integral

$$\frac{1}{2\pi(n - m)} \int_K f(x + t) \frac{\sin \dfrac{n + m}{2} t \sin \dfrac{n - m}{2} t}{\sin^2 (t/2)} \, dt.$$

9. Prove that

$$\| V_{m,n} \| = \frac{4}{\pi^2} \log \frac{n}{n-m} + O(1).$$

If $m = m(n) < n$, derive from this and Problem 7 the conditions of convergence $V_{m,n}(f) \to f$, $n \to \infty$, for all $f \in C^*$.

10. Let $\lambda_k^{(n)}$, $n, k = 1, 2, \cdots$ be a double sequence for which $\sum\limits_{k=1}^{\infty} |\lambda_k^{(n)}| < +\infty$, $n = 1, 2, \cdots$, and let $L_n(f, x) = \frac{1}{2} a_0 + \sum\limits_{k=1}^{\infty} \lambda_k^{(n)} (a_k \cos kx + b_k \sin kx)$, where a_k, b_k are the Fourier coefficients of $f \in C^*$. Assume that (*) $\lim\limits_{n\to\infty} n^r(1 - \lambda_k^{(n)}) = \phi(k)$, $k = 1, 2, \cdots$, with some $r > 0$, $\phi(k) > 0$. Then $\| f - L_n(f) \| = o(n^{-r})$ implies that f is a constant.

11. If the condition (*) is satisfied, then $\| f - L_n(f) \| = O(n^{-r})$ implies that the arithmetic mean of the series $f_r(x) = \sum\limits_{k=1}^{\infty} k^r(a_k \cos kx + b_k \sin kx)$ satisfy $\| \sigma_N(f_r) \| = O(1)$ for $N \to \infty$ (Sunouchi and Watari). [*Hint:* Consider $\sigma_N(f - L_n(f))$.]

12. The Jackson integral 4.2(3) has the form

$$J_n(f, x) = \sum_{k=0}^{2n-2} \lambda_k^{(n)}(a_k \cos kx + b_k \sin kx),$$

where $\lambda_k^{(n)} = 1 - \frac{3}{2}(k/n)^2 + O(n^{-3})$ for each k as $n \to \infty$.

13. The saturation class of the operator $J_n(f)$ consists of all functions $f \in C^*$ with $f' \in \text{Lip } 1$, and the optimal order of approximation is $\psi(n) = n^{-2}$. (*Hint:* Use Problems 10-12 and Problem 4 of Chapter 3.)

[8]

Approximation of Classes of Functions

1. Introduction

In Chapters 4 and 5 we have obtained, for functions f of certain function classes, upper bounds for the degree of approximation of f by trigonometric or algebraic polynomials. All these results contain an unspecified constant M. If we pursue the proofs a little further, it is easy each time to find a concrete numerical value for M. We ask now for the *best possible constants* M.

There is no hope of answering this question for an arbitrary f. Therefore, we shall turn our attention to *classes of functions*. Let $\Re$ be a subset of C^*. We define *the degree of approximation of $\Re$ by trigonometric polynomials of degree n* by the formula

$$E_n^*(\Re) = \sup_{f \in \Re} E_n^*(f). \tag{1}$$

We see that the "worst functions" of $\Re$ determine $E_n^*(\Re)$. A little more generally, let $\Re$ and $\Re'$ be two subsets of a metric space. Then

$$E_{\Re'}(f) = \inf_{g \in \Re'} \rho(f, g)$$

is the degree of approximation of f by the elements $g \in \Re'$, and

$$E_{\Re'}(\Re) = \sup_{f \in \Re} E_{\Re'}(f) \tag{2}$$

is the *degree of approximation of $\Re$ by $\Re'$*, or the *deviation of $\Re$ from $\Re'$*. Obviously, $E_n^*(\Re)$ is equal to $E_{\Re'}(\Re)$ if $\Re'$ is the space of all trigonometric polynomials of degree n, and $\Re$ is a subset of C^*.

The purpose of this chapter is to determine the exact value of $E_n^*(\Re)$ for the classes W_p^*, $p = 1, 2, \cdots$ (see Chapter 3, Sec. 7), the classes Lip α, $0 < \alpha \leqslant 1$ and some classes of analytic functions. It is instructive to see how the problem of *uniform* approximation can be related to the approximation *in the L^1 norm*.

The space $L^1[A]$, $A = [a, b]$, or $A = K$, consists of all Lebesgue integrable real functions f on A, with the norm $\| f \|_1 = \int_A | f(x) | \, dx$. If $A = K$, $f \in L^1$, the degree of approximation of f by trigonometric polynomials in this norm will be denoted by $E_n^*(f)_1$. The usefulness of the L^1 approximation for our purposes has its root in a convolution formula [see 3(13)],

$$f(x) = \frac{a_0}{2} + \int_K G(x - t)\, h(t) \, dt,$$

111

which is available for all $f \in W_p^*$. Here, G is some fixed function, while h depends on f, and satisfies $|h(t)| \leqslant 1$. We have the *duality* of the L^1- and the C^*-approximations: If $S_n(u)$ is a trigonometric polynomial that approximates $G(u)$ with an error $\leqslant \epsilon$ *in the L^1 norm*, $\int_K |G - S_n| \, du \leqslant \epsilon$, then

$$T_n(x) = \tfrac{1}{2} a_0 + \int_K S_n(x - t) \, h(t) \, dt$$

is another trigonometric polynomial, which approximates $f(x)$ *uniformly*: $|f(x) - T_n(x)| \leqslant \epsilon$.

All our results will be about *trigonometric approximation*; no comparable results for algebraic approximation are yet known (except for some classes of analytic functions).

2. Approximation in the Space L^1

The theorems of this section deal with exceptional cases, when the polynomials of best approximation (in the L^1 norm) can be determined explicitly by means of simple algebraic operations and integration.

THEOREM 1. Let $\phi_1(x), \cdots, \phi_n(x)$ be a system of integrable functions on $A = [a, b]$ or $A = K$, and let $f(x)$ also be an integrable function. Then $P(x) = \sum_1^n a_k \phi_k(x)$ is a polynomial of best approximation of f if

$$\int_A \phi_k(x) \operatorname{sign} [f(x) - P(x)] \, dx = 0, \qquad k = 1, \cdots, n. \tag{1}$$

If $f(x) - P(x)$ vanishes only on a set of measure zero, this condition is also necessary.

Proof. Let the condition (1) be satisfied. If Q is a polynomial in the ϕ_k, then $\int_A (Q - P) \operatorname{sign} (f - P) \, dx = 0$; hence,

$$\int_A |f - Q| \, dx \geqslant \int_A (f - Q) \operatorname{sign} (f - P) \, dx$$

$$= \int_A (f - P) \operatorname{sign} (f - P) \, dx$$

$$= \int_A |f - P| \, dx,$$

so that condition (1) is sufficient. That the condition is necessary is a little harder to prove (and is less useful). Assume that, for some k, $1 \leqslant k \leqslant n$,

$$\int_A \phi_k \operatorname{sign} (f - P) \, dx = \rho \neq 0.$$

At a point x, where $f(x) - P(x) \neq 0$, that is, almost everywhere, we have

$$\text{sign}\,[f(x) - P(x) - \lambda\phi_k(x)] \to \text{sign}\,[f(x) - P(x)] \quad\text{if}\quad \lambda \to 0.$$

By Lebesgue's convergence theorem, we have, therefore,

$$\int_A \phi_k \,\text{sign}\,[f - P - \lambda\phi_k]\,dx \to \rho, \qquad \lambda \to 0.$$

Hence, it is possible to select λ in such a way that

$$\lambda \int_A \phi_k \,\text{sign}\,[f - P - \lambda\phi_k]\,dx > 0.$$

If we put $Q = P + \lambda\phi_k$, we shall have

$$\int_A |f - P|\,dx \geqslant \int_A (f - P)\,\text{sign}\,(f - Q)\,dx$$

$$> \int_A (f - P - \lambda\phi_k)\,\text{sign}\,(f - Q)\,dx$$

$$= \int_A |f - Q|\,dx.$$

Thus, P is *not* a polynomial of best approximation.   ▌

Often Theorem 1 is applied in the following situation: Assume that we have found points $x_1 < x_2 < \cdots < x_m$ of A and a polynomial P with the following properties:

(a) The difference $f(x) - P(x)$ changes sign at the x_k, and retains a constant sign between these points.

(b) If $\Phi(x)$ is the function $\Phi(x) = \pm 1$ on A, which changes sign precisely at the points x_k,

$$\int_A \phi_k \Phi\,dx = 0, \qquad k = 1, \cdots, m. \tag{2}$$

Then P is a polynomial of best approximation to f in the L^1 norm, and the degree of approximation of f by the polynomials in the ϕ_k is

$$E_n(f)_1 = \left| \int_A f(x)\,\Phi(x)\,dx \right|. \tag{3}$$

The *proof* is obvious: We have $\Phi(x) = \epsilon\,\text{sign}\,[f(x) - P(x)]$ a.e., where

$\epsilon = 1$ or $\epsilon = -1$; hence, (2) gives condition (1). Moreover,

$$\int_A |f - P| \, dx = \int_A (f - P) \operatorname{sign}(f - P) \, dx$$

$$= \int_A f \operatorname{sign}(f - P) \, dx$$

$$= \left| \int_A f\Phi \, dx \right|.$$

We apply this to trigonometric approximation. The next theorem has two variants: for an even function G and for an odd function H. We shall say that an even [or odd] function G [or H] on K *satisfies the condition* A_n if it is integrable on K and continuous and indefinitely differentiable on K, except perhaps at $x = 0$. For each selection of real coefficients λ_k, $k = 0, \cdots, n - 1$, the difference

$$G(x) - \sum_{k=0}^{n-1} \lambda_k \cos kx \left[\text{or } H(x) - \sum_{k=1}^{n-1} \lambda_k \sin kx \right]$$

has at most n [or $n - 1$] zeros (counting their multiplicity) on the *open* interval $(0, \pi)$.

THEOREM 2.   Let G [or H] be a function on K that satisfies the condition A_n. Let $T_{n-1}(x)$ be the even [or odd] trigonometric polynomial of degree $n - 1$, which interpolates G [or H] at the n [or $n - 1$] zeros of $\cos nx$ [or $\sin nx$] in the open interval $(0, \pi)$. Then T_{n-1} is a polynomial of best approximation for G [or H] in the L^1 norm, and

$$E_{n-1}^*(G)_1 = \left| \int_K G(x) \operatorname{sign} \cos nx \, dx \right|,$$

$$E_{n-1}^*(H)_1 = \left| \int_K H(x) \operatorname{sign} \sin nx \, dx \right|.$$

(4)

Proof. If $\tilde{T}_{n-1}(x)$ is a polynomial of best approximation for $G(x)$ on K, then also $\tilde{T}_{n-1}(-x)$ and $T_{n-1}(x) = \frac{1}{2}[\tilde{T}_{n-1}(x) + \tilde{T}_{n-1}(-x)]$ have this property. Hence, G has an *even* polynomial of best approximation. [Actually (Sec. 8, Note 1) G has a *unique* polynomial of best approximation.] Similarly, H has an *odd* polynomial of best approximation. Thus, we may consider the approximation of G [or H] by even [or odd] trigonometric polynomials.

Let the system $\phi_1, \cdots, \phi_n$ of Theorem 1 be 1, $\cos x, \cdots, \cos(n-1)x$ [or $\sin x, \cdots, \sin(n-1)x$]. Let x_k, $k = 1, \cdots, n$ [or $k = 1, \cdots, n-1$] be the zeros of $\cos nx$ [or $\sin nx$] in $(0, \pi)$. The interpolating polynomial $T_{n-1}(x) = \sum_{k=0}^{n-1} \lambda_k \cos kx$ [or $= \sum_{k=1}^{n-1} \lambda_k \sin kx$] exists, since $\phi_1, \cdots, \phi_n$ is a Chebyshev system on $[0, \pi]$. Because of the condition A_n, the difference $G - T_{n-1}$ [or $H - T_{n-1}$] has the x_k as its only zeros, and each x_k is a simple zero. Thus condition (a) is satisfied. In condition (b), we shall take $\Phi(x) = \text{sign} \cos nx$ [or $= \text{sign} \sin nx$]. Relation (2) follows from

$$\int_K \Phi(x) \cos kx \, dx = \int_K \Phi(x) \sin kx \, dx = 0, \qquad k = 0, \cdots, n-1. \quad (5)$$

This is obvious if $k = 0$. For $1 \leqslant k \leqslant n-1$, see 4.3(9). ∎

A very similar theorem is the following:

THEOREM 3 (Bernstein [2, vol. I, pp. 330-332]). Let $f^{(n+1)}(x) > 0$ in $(-1, +1)$. Then the algebraic polynomial of best approximation to $f(x)$ in the L^1 norm is the polynomial $P_n(x)$, which interpolates $f(x)$ at the $n+1$ points $x_k = \cos(k\pi/(n+2))$, $k = 1, \cdots, n+1$.

Proof. Let $P_n(x)$ interpolate f at the points x_k. Assume that $f(x) - P_n(x) = g(x)$ has $n+2$ zeros (counting their multiplicity). Then, by Rolle's theorem, $g^{(n+1)}(x)$ has a zero in $(-1, +1)$. But $g^{(n+1)}(x) = f^{(n+1)}(x) > 0$. Thus, g has $n+1$ simple zeros at the x_k and changes sign at exactly these points. Condition (2) takes the form

$$\int_{-1}^{+1} x^m \Phi(x) \, dx = 0, \qquad m = 0, 1, \cdots, n,$$

or upon substituting $x = \cos t$,

$$\int_0^\pi \cos^m t \sin t \Psi(t) \, dt = 0, \qquad (6)$$

where $\Psi(t) = \pm 1$ is the function that changes sign exactly at the points $t_k = k\pi/(n+2)$, $k = 1, \cdots, n+1$. This is equivalent to

$$\int_K \cos^m t \sin t \Psi(t) \, dt = 0, \qquad \Psi(t) = \text{sign} \sin(n+2)t.$$

But $\cos^m t \sin t$ for $m = 0, 1, \cdots, n$ are trigonometric polynomials of degrees $\leqslant n+1$, so that (6) is satisfied because of 4.3(9). ∎

A function $f(x)$ is called *absolutely monotone* in $[-1, +1]$ if all derivatives $f^{(n)}(x)$ exist and all have the same sign in this interval. We see that for absolutely monotone functions, all polynomials of best L^1 approximation are easily described.

3. The Degree of Approximation of the Classes W_p^*

We now deal with uniform approximation. The main theorem is

THEOREM 4 (Favard [49]). The degree of approximation of the classes W_p^*, $p = 1, 2, \cdots$ is given by

$$E_{n-1}^*(W_p^*) = K_p \frac{1}{n^p}, \qquad n = 1, 2, \cdots, \quad (1)$$

where

$$K_p = \begin{cases} \dfrac{4}{\pi} \displaystyle\sum_{m=0}^{\infty} \dfrac{1}{(2m+1)^{p+1}} & \text{if } p \text{ is odd,} \\[3mm] \dfrac{4}{\pi} \displaystyle\sum_{m=0}^{\infty} \dfrac{(-1)^m}{(2m+1)^{p+1}} & \text{if } p \text{ is even.} \end{cases} \tag{2}$$

For each $n, p = 1, 2, \cdots$, there exists an extremal function $f_{np} \in W_p^*$, given by (16), for which

$$E_{n-1}^*(W_p^*) = \|f_{np}\| = |f_{np}(0)|; \tag{3}$$

this function has $2n$ points of maxima and minima on K, where it takes the values $\pm K_p n^{-p}$ with alternating signs.

The K_p with odd subscripts p can be shown to be rational expressions in π. For example, using a known formula, we have

$$K_1 = \frac{4}{\pi} \sum_{m=0}^{\infty} \frac{1}{(2m+1)^2} = \frac{4}{\pi} \cdot \frac{\pi^2}{8} = \frac{1}{2} \pi. \tag{4}$$

It will be convenient to prove a slightly more general theorem.

THEOREM 5.   Let G [or H] be a given function on K, which satisfies condition A_n. By $\Re(G)$ [or $\Re(H)$], we denote the class of all functions f that are representable in the form

$$f(x) = \int_K G(x-t) h(t)\, dt \left[\text{or } f(x) = \int_K H(x-t) h(t)\, dt \right] \tag{5}$$

with some measurable h with $|h(t)| \leqslant 1$ a.e. Let

$$\left. \begin{aligned} f_{nG}(x) &= \int_K G(x-t) \operatorname{sign} \cos nt\, dt \\[2mm] f_{nH}(x) &= \int_K H(x-t) \operatorname{sign} \sin nt\, dt \end{aligned} \right\} \quad n = 1, 2, \cdots. \tag{6}$$

Then, for the uniform approximation,

$$\begin{aligned} E_{n-1}^*(\Re(G)) &= \|f_{nG}\| = |f_{nG}(0)|, \\[2mm] E_{n-1}^*(\Re(H)) &= \|f_{nH}\| = |f_{nH}(0)|. \end{aligned} \tag{7}$$

Proof of Theorem 5.   We shall give the proof for a function G only; the case of a function H is quite similar. Clearly, $f_n = f_{nG} \in \Re(G)$. We have

$$f_n\left(x + \frac{\pi}{n}\right) = \int_K G(t) \operatorname{sign} \cos n\left(x + \frac{\pi}{n} - t\right) dt = -f_n(x).$$

If $|f_n(x)|$ attains its maximum at some point x_0, then also all points $x_k = x_0 + k\pi/n$, $k = 0, \pm 1, \cdots$, give to $|f_n(x)|$ its maximum, and the signs of $f_n(x)$ at the x_k alternate. (In a moment we shall see that $x_0 = 0$ is a possible choice.) There are $2n$ distinct points x_k on K. By Chebyshev's theorem, the constant zero is the trigonometric polynomial of degree $n - 1$ of best approximation to f_n. Therefore,

$$E_{n-1}^*(\Re(G)) \geqslant E_{n-1}^*(f_n) = \|f_n\| \geqslant |f_n(0)|. \tag{8}$$

Next we shall estimate $E_{n-1}^*(\Re(G))$ from above. Let S_{n-1} be a polynomial of best approximation for G in the L^1 norm, and let $f \in \Re(G)$ be any function of the form (5). Then

$$T_{n-1}(x) = \int_K S_{n-1}(x - t)\, h(t)\, dt$$

is also a trigonometric polynomial of degree $n - 1$. Using 2(4), we obtain

$$E_{n-1}^*(f) \leqslant \|f - T_{n-1}\| = \max_x |f(x) - T_{n-1}(x)|$$

$$\leqslant \max_x \int_K |G(x - t) - S_{n-1}(x - t)| \cdot |h(t)|\, dt \tag{9}$$

$$\leqslant E_{n-1}^*(G)_1$$

$$= \left| \int_K G(t)\, \text{sign} \cos nt\, dt \right| = |f_n(0)|.$$

This proves that $E_{n-1}^*(\Re(G)) \leqslant |f_n(0)|$. █

Proof of Theorem 4. We consider the functions

$$D_p(x) = \frac{1}{\pi} \sum_{k=1}^{\infty} \frac{\cos(kx - p\pi/2)}{k^p}, \qquad p = 1, 2, \cdots. \tag{10}$$

For $p \geqslant 2$, the series (10) converges uniformly on K. If $p = 1$, (10) converges on K with uniformly bounded partial sums (see [24, p: 29]) and uniformly in $\epsilon < x < \pi - \epsilon$ for $0 < \epsilon < \pi/2$. In this case, (10) is the Fourier series of an elementary function:

$$D_1(x) = \frac{\pi - x}{2\pi} = \frac{1}{\pi} \sum_{k=1}^{\infty} \frac{\sin kx}{k}, \qquad 0 < x < 2\pi.$$

If we define $D_0(x) = -(1/2\pi)$, then we have the relation

$$D_p'(x) = D_{p-1}(x), \qquad p = 1, 2, 3, \cdots, x \in K \quad (\text{if } p = 1, 2, x \neq 0). \tag{11}$$

We see that the D_p are continuous and differentiable on K, except possibly for $x = 0$. A function D_p is even or odd according to whether p is even or odd.

LEMMA 1.   For $p, n = 1, 2, \cdots$, the functions D_p satisfy the condition A_n.

We first show that if D_p, $p \geqslant 2$, does not satisfy A_n, then D_{p-1} does not. Indeed, if p is even, the hypothesis implies the existence of an even trigonometric polynomial T_{n-1}, for which $D_p - T_{n-1}$ has more than n zeros (counting their multiplicity) on $(0, \pi)$. By Rolle's theorem and (11), $D_{p-1} - T'_{n-1}$ has more than $n - 1$ zeros on $(0, \pi)$; that is, D_{p-1} does not satisfy A_n. The same argument applies when p is odd (then $p \geqslant 3$), if one takes into account that each odd continuous function on K has zeros $0, \pi$.

Thus, if D_p, $p \geqslant 1$, does not satisfy A_n, then neither does D_1. But this implies that for some odd trigonometric polynomial T_{n-1}, $D_1 - T_{n-1}$ has at least n zeros in $(0, \pi)$. But π is also a zero. By Rolle's theorem, the derivative $-(1/2\pi) - T'_{n-1}$ has at least n zeros in $(0, \pi)$: hence it has $2n$ zeros on K, and this is impossible.   ▮

For a function $f \in W_p^*$, let

$$f(x) = \frac{a_0}{2} + \sum_{k=1}^{\infty} (a_k \cos kx + b_k \sin kx)$$

be the (convergent) Fourier series of f. Let W_{p0}^* denote the set of all $f \in W_p^*$ with mean-value zero; that is, with $a_0 = 0$. The degrees of approximation of f and of $f - (a_0/2)$ are the same; hence $E_{n-1}^*(W_p^*) = E_{n-1}^*(W_{p0}^*)$, $n = 1, 2, \cdots$, and we can replace W_p^* by W_{p0}^* in the statement of Theorem 4.

We now show that the functions $f \in W_{p0}^*$ have a representation of the form (5). By integration by parts, we obtain

$$a_k \cos kx + b_k \sin kx = \frac{1}{\pi} \int_K f(t) \cos k(x - t) \, dt$$

$$= \frac{1}{\pi} \int_K f'(t) \frac{\cos (k(x - t) - \pi/2)}{k} \, dt;$$

continuing in the same way, we have

$$a_k \cos kx + b_k \sin kx = \frac{1}{\pi} \int_K f^{(p)}(t) \frac{\cos (k(x - t) - p\pi/2)}{k^p} \, dt. \qquad (12)$$

Because of the uniform or bounded convergence of the series (10), we obtain

$$f(x) = \int_K D_p(x - t) h(t) \, dt, \qquad f \in W_{p0}^*, \qquad (13)$$

where $h(t) = f^{(p)}(t)$ and $| h(t) | \leqslant 1$ a.e. We see that $W_{p0}^* \subset \Re(D_p)$.

Conversely, if a bounded function h of mean value zero is given, it has a unique pth indefinite integral on K with mean-value zero (Chapter 1, Sec. 1). It is now clear that this integral is given by (13).

In particular, let f_{np} be the extremal function of $\Re(D_p)$. It is given by (6), with G [or H] equal to D_p; that is, by (13) with

$$h(x) = \begin{cases} \text{sign sin } nx, & \text{if} \quad p \text{ is odd,} \\ \text{sign cos } nx, & \text{if} \quad p \text{ is even.} \end{cases} \tag{14}$$

Hence, (3) follows from (7) and the inclusion $W_{p0}^* \subset \Re(D_p)$.

It remains to compute $f_{np}(x)$ and $|f_{np}(0)|$. We start with the Fourier series expansion

$$\frac{4}{\pi} \sum_{m=0}^{\infty} \frac{\sin (2m + 1) x}{2m + 1} = \begin{cases} + 1, & \text{if} \quad 0 < x < \pi, \\ - 1, & \text{if} \quad -\pi < x < 0, \\ 0, & \text{if} \quad x = 0, \pi. \end{cases}$$

The right-hand side is equal to sign sin x on K. Replacing x by nx or by $nx + \frac{1}{2}\pi$, respectively, we have for the function (14) on K,

$$h(x) = \begin{cases} \dfrac{4}{\pi} \displaystyle\sum_{m=0}^{\infty} \dfrac{\sin n(2m + 1) x}{2m + 1}, & \text{if } p \text{ is odd,} \\[4mm] \dfrac{4}{\pi} \displaystyle\sum_{m=0}^{\infty} (-1)^m \dfrac{\cos n(2m + 1) x}{2m + 1}, & \text{if } p \text{ is even.} \end{cases} \tag{15}$$

Relation (12) tells us that for $f \in W_{p0}^*$, the Fourier coefficients a_k, b_k of f and α_k, β_k of $f^{(p)}$ are connected by

$$\left. \begin{aligned} a_k &= (-1)^{(p+1)/2} k^{-p} \beta_k \\ b_k &= (-1)^{(p-1)/2} k^{-p} \alpha_k \end{aligned} \right\} \quad \text{if } p \text{ is odd,}$$

and by

$$\left. \begin{aligned} a_k &= (-1)^{p/2} k^{-p} \alpha_k \\ b_k &= (-1)^{p/2} k^{-p} \beta_k \end{aligned} \right\} \quad \text{if } p \text{ is even.}$$

Therefore

$$f_{np}(x) = \begin{cases} (-1)^{(p+1)/2} \dfrac{4}{\pi n^p} \displaystyle\sum_{m=0}^{\infty} \dfrac{\cos n(2m + 1) x}{(2m + 1)^{p+1}}, & \text{if } p \text{ is odd,} \\[4mm] (-1)^{p/2} \dfrac{4}{\pi n^p} \displaystyle\sum_{m=0}^{\infty} (-1)^m \dfrac{\cos n(2m + 1) x}{(2m + 1)^{p+1}}, & \text{if } p \text{ is even.} \end{cases} \tag{16}$$

From this we obtain the values (1).   ▮

A corollary of Favard's theorem is that

$$E_{n-1}^*(W_p^*) \leqslant \frac{\pi}{2} n^{-p}, \quad n = 1, 2, \cdots, \tag{17}$$

while from Jackson's theorem it follows only that the left-hand side does not exceed $C_p n^{-p}$, with some C_p depending on p. Indeed, $K_p \leqslant K_1 = \frac{1}{2}\pi$ for odd p, $K_p \leqslant 4/\pi$ for even p.

If L is a linear operator that maps C^* into itself, and $\Re \subset C^*$, we call

$$E(\Re; L) = \sup_{f \in \Re} \| f - L(f) \| \qquad (18)$$

the *degree of approximation of* $\Re$ *by the operator* L. The following problem is interesting: If a set $\Re \subset C^*$ and an integer $n = 0, 1, \cdots$ are given, does there exist a linear polynomial operator L_n of degree n for which $E(\Re; L_n) = E_n^*(\Re)$? An operator of this kind will be called *an optimal operator for* $\Re$.

THEOREM 6.   The classes W_p^*, $p = 1, 2, \cdots$, have optimal operators for each $n = 0, 1, \cdots$.

Proof.   Let S_{n-1} (p. 117) be the polynomial that approximates best D_p in the L^1 norm. We put

$$L_{n-1}(f, x) = \frac{1}{2}a_0 + \int_K S_{n-1}(x - t) f^{(p)}(t)\, dt$$

$$= \frac{1}{2}a_0 + \int_K S_{n-1}^{(p)}(x - t) f(t)\, dt. \qquad (19)$$

Inequalities (9) show that

$$\| f - L_{n-1}(f) \| \leqslant E_{n-1}^*(W_p^*), \qquad f \in W_p^*. \qquad \blacksquare$$

4.   Distance Matrices

Let $\omega(h)$ be a modulus of continuity, that is, a function defined for $0 \leqslant h \leqslant l$, which satisfies $\omega(0) = 0$, $\omega(h) \geqslant 0$, and is continuous, increasing, and subadditive (see Chapter 3, Sec. 5). Then Λ_ω is the set of all real functions f defined on $[a, b]$, or K, which satisfy $\omega(f, h) \leqslant \omega(h)$ for $0 \leqslant h \leqslant l$ (we take $l = b - a$ or, correspondingly, $l = \pi$). In the next section we shall find the exact value of the degree of approximation, $E_n^*(\Lambda_\omega)$, of the class Λ_ω by trigonometric polynomials of degree n, for the case when ω is concave. This includes all functions $\omega(h) = Mh^\alpha$, $0 < \alpha \leqslant 1$. The main results, Theorems 9, 10, and 11, are due to Korneĭchuk [66, 68, 69], but the proof of Theorem 9 via Theorem 7 is Timan's [97].

Let $X = (x_1, \cdots, x_n)$ be an n-tuple of points of a metric space, not necessarily distinct. The matrix $A = (a_{ij})$ with elements $a_{ij} = \rho(x_i, x_j)$ has the following properties:

$$\left. \begin{array}{ll} a_{ij} \geqslant 0 \quad \text{and} \quad a_{ii} = 0 & \\ a_{ij} = a_{ji} & i, j, k = 1, \cdots, n. \\ a_{ik} \leqslant a_{ij} + a_{jk} & \end{array} \right. \qquad \begin{array}{l}(1)\\(2)\\(3)\end{array}$$

As a consequence, we have

$$| a_{ik} - a_{jk} | \leqslant a_{ij} . \tag{4}$$

A matrix A that satisfies the three conditions (1) to (3) will be called a *distance matrix*. (This name is justified by the fact, which is easy to verify, that for each distance matrix A, there is a metric space containing an n-tuple of points x_i, for which $\rho(x_i, x_j) = a_{ij}$.) We shall say that an n-tuple X is *dominated* by the distance matrix A, if $\rho(x_i, x_j) \leqslant a_{ij}$ for $i, j = 1, \cdots, n$.

We are interested in the case when $x_1, \cdots, x_n$ are points of the real line R. For n-tuples $X = (x_1, \cdots, x_n)$, $Y = (y_1, \cdots, y_n)$ on R, we define a distance by the formula

$$d(X, Y) = \max_i | x_i - y_i | . \tag{5}$$

Our problem will be to approximate an arbitrary n-tuple X with x_i in R by an n-tuple dominated by a given distance matrix B. Let B' denote the set of all Y dominated by B. For the degree of approximation of X by B' (see Sec. 1) we have

THEOREM 7. *If X is dominated by the distance matrix A, then*

$$E_{B'}(X) \leqslant \tfrac{1}{2} \max_{i,j} (a_{ij} - b_{ij}). \tag{6}$$

Proof. Let d be the right-hand side of (6). We have to find an n-tuple $Y \in B'$ for which $d(X, Y) \leqslant d$.

We define an n-tuple Z by putting

$$z_i = \min_{k \leqslant n} \{x_k + b_{ki}\}. \tag{7}$$

Then Z is dominated by B: $| z_i - z_j | \leqslant b_{ij}$. If, for example, $z_i \leqslant z_j$, then we select a k with $z_i = x_k + b_{ik}$ and have

$$| z_i - z_j | = z_j - z_i$$

$$\leqslant \{x_k + b_{kj}\} - \{x_k + b_{ki}\}$$

$$= b_{kj} - b_{ki}$$

$$\leqslant b_{ij} .$$

Next we prove that

$$0 \leqslant x_i - z_i \leqslant 2d. \tag{8}$$

The left-hand inequality holds, since $z_i \leqslant x_i$. On the other hand, $z_i = x_k + b_{ki}$ for some $1 \leqslant k \leqslant i$; hence,

$$x_i - z_i = x_i - x_k - b_{ik} \leqslant a_{ik} - b_{ik} \leqslant 2d.$$

We put $y_i = z_i + d, i = 1, \cdots, n$; then the n-tuple $Y = (y_1, \cdots, y_n)$ is dominated by B and $d(X, Y) \leqslant d$. ∎

5. Approximation of the Classes Λ_ω

As an application of Theorem 7 we have the following result:

THEOREM 8. If $A = [a, b]$ or $A = K$, $l = b - a$ or $l = \pi$, correspondingly, and if $\omega(h)$, $w(h)$ are two moduli of continuity defined for $0 \leqslant h \leqslant l$, then

$$E_{\Lambda_w}(\Lambda_\omega) = \tfrac{1}{2} \max_{0 \leqslant h \leqslant l} [\omega(h) - w(h)]. \tag{1}$$

Proof. (a) Let d be the right-hand side of (1). We first prove that $E_{\Lambda_w}(\Lambda_\omega) \leqslant d$. For this purpose, we shall, for a given function $f \in \Lambda_\omega$; find a function $g \in \Lambda_w$ for which $\| f - g \| \leqslant d$.

Let $D = \{x_1, \cdots, x_n, \cdots\}$ be a countable, dense set in A, and let $D_n = \{x_1, \cdots, x_n\}$, $n = 1, 2, \cdots$. The n-tuple $X = (f(x_1), \cdots, f(x_n))$ is dominated by the matrix $(\omega(|\, x_i - x_j\,|))$. Let B be the distance matrix with elements $w(|\, x_i - x_j\,|)$. According to Theorem 7, there exists an n-tuple

$$Y = (y_1, \cdots, y_n) \in B',$$

which approximates X with an error not exceeding d: $|\, f(x_i) - y_i\,| \leqslant d$, $i = 1, \cdots, n$. We define a function g_n on D_n by putting $g_n(x_i) = y_i, i = 1, \cdots, n$. The functions g_n are uniformly bounded: $|\, g_n(x)\,| \leqslant \| f \| + d$ holds for each x where g_n is defined. We can therefore extract a subsequence g_{n_k}, $k = 1, 2, \cdots$, which converges at each point x_i of D: $\lim\limits_{k \to \infty} g_{n_k}(x_i) = g(x_i), i = 1, 2, \cdots$. The function g obviously satisfies $|\, g(x_i) - g(x_j)\,| \leqslant w(|\, x_i - x_j\,|), i, j = 1, 2, \cdots$, on D. Also, $|\, f(x_i) - g(x_i)\,| \leqslant d, i = 1, 2, \cdots$. Since g is uniformly continuous on the dense set D, we can extend it to a continuous function g on A, which has all the required properties.

(b) To complete the proof, we have to exhibit a function $f_0 \in \Lambda_\omega$, which cannot be approximated by the $g \in \Lambda_w$ with an error $< d$.

Let $f_0(x) = \omega(x - a)$ for $A = [a, b]$. If $A = K$, we put $f_0(x) = \omega(|\, x\,|)$ for $-\pi \leqslant x \leqslant \pi$. Then (compare p. 43), $f_0 \in \Lambda_\omega$.

We select an h, $0 \leqslant h \leqslant l$ for which $d = \tfrac{1}{2} [\omega(h) - w(h)]$. For each $g \in \Lambda_w$, we have (we put $a = 0$ if $A = K$)

$$\begin{aligned}
2 \| f_0 - g \| &\geqslant |\, f_0(a + h) - g(a + h) + g(a)\,| \\
&\geqslant f_0(a + h) - |\, g(a + h) - g(a)\,| \\
&\geqslant \omega(h) - w(h) = 2d. \quad ∎
\end{aligned}$$

LEMMA 2. Let ω be a concave modulus of continuity. Then, for each h, $0 < h \leqslant l$, there is an $M \geqslant 0$ for which

$$\max_{0 \leqslant x \leqslant l} [\omega(x) - Mx] = \omega(h) - Mh. \tag{2}$$

Proof. We use the following facts about concave, increasing functions $\omega(x)$ (see [23, p. 94]). At each point h, $0 < h \leqslant l$, the function ω has a finite right $\omega'_+(h)$ and a finite left $\omega'_-(h)$ derivative, and $\omega'_+(h) \leqslant \omega'_-(h)$. Moreover, if M is contained between the last two numbers, then the curve $y = \omega(x)$ lies below the straight line with slope M, which passes through the point $(h, \omega(h))$. This gives

$$\omega(x) \leqslant M(x - h) + \omega(h), \qquad 0 \leqslant x \leqslant l. \qquad \blacksquare$$

The main result of this section is the following

THEOREM 9. Let ω be a concave modulus of continuity. Then

$$E^*_{n-1}(\Lambda_\omega) = \max_{f \in \Lambda_\omega} E^*_{n-1}(f) = \frac{1}{2}\,\omega\left(\frac{\pi}{n}\right), \qquad n = 1, 2, \cdots. \tag{3}$$

The maximum in (3) is attained for the function $f_n \in \Lambda_\omega$, given by (6) below.

Proof. Let $M = M(n) \geqslant 0$ denote a value of M for which (2) holds, with $h = \pi/n$. According to Theorem 8 and Lemma 2, for each $f \in \Lambda_\omega$ there is a function $g \in \mathrm{Lip}_M 1$ such that

$$\|f - g\| \leqslant \tfrac{1}{2} \max_{0 \leqslant u \leqslant \pi} [\omega(u) - Mu] = \tfrac{1}{2}\,\omega\left(\frac{\pi}{n}\right) - \frac{\pi M}{2n}. \tag{4}$$

By Theorem 4 and 3(4), each function of the class $W^*_1 = \mathrm{Lip}_1 1$ can be approximated by a trigonometric polynomial of degree $n - 1$, with an error not exceeding $\tfrac{1}{2}\,\pi/n$. Since $(g/M) \in W^*_1$, for some T_{n-1} we have

$$\|g - T_{n-1}\| \leqslant \frac{\pi M}{2n}. \tag{5}$$

From (4) and (5), $\|f - T_{n-1}\| \leqslant \tfrac{1}{2}\omega(\pi/n)$. It follows that $E^*_{n-1}(\Lambda_\omega)$ has the same upper bound.

Let f_n be the odd function on K of period $2\pi/n$, given by

$$f_n(x) = \begin{cases} \tfrac{1}{2}\,\omega(2x), & 0 \leqslant x \leqslant \dfrac{\pi}{2n}, \\[2mm] \tfrac{1}{2}\,\omega\left(\dfrac{2\pi}{n} - 2x\right), & \dfrac{\pi}{2n} \leqslant x \leqslant \dfrac{\pi}{n}. \end{cases} \tag{6}$$

By 3.5(9), $f_n \in \Lambda_\omega$. On K, f_n has exactly $2n$ points of maxima and minima, where it takes values $\pm \frac{1}{2} \omega(\pi/n)$ with alternating signs. By Chebyshev's theorem, the constant zero is the trigonometric polynomial of degree $n - 1$ of best approximation for f_n . Hence,

$$E_{n-1}^*(\Lambda_\omega) \geqslant E_{n-1}^*(f_n) \geqslant \frac{1}{2} \omega\left(\frac{\pi}{n}\right).$$

6. Arbitrary Moduli of Continuity; Approximation by Operators

For a nonconcave modulus of continuity ω, Theorem 9 is not always true. Indeed, we have

LEMMA 3. For each $n = 1, 2, \cdots$, and each $\epsilon > 0$, there is a function $g \in C^*$ with

$$E_{n-1}^*(g) \geqslant \left(\frac{2n - 1}{2n} - \epsilon\right) \omega\left(g, \frac{\pi}{n}\right). \tag{1}$$

Proof. The points $x_k = k\pi/n, \ k = 0, 1, \cdots, n$ are π/n apart in $[0, \pi]$, but the points $y_k = k\pi/n - (n - k)\,\delta, \ k = 1, \cdots, n$ are $(\pi/n) + \delta$ apart. We take $\delta > 0$ and small; then $0 < y_1 - \delta < y_1 < \cdots < y_n = \pi$. Now

$$T_{n-1}(x) = \frac{1}{n} D_{n-1}(x) = \frac{1}{n} \frac{\sin (n - \frac{1}{2}) x}{2 \sin (x/2)}$$

is an even trigonometric polynomial, and

$$T_{n-1}(0) = \frac{2n - 1}{2n}, \qquad T_{n-1}(x_k) = \frac{(-1)^{k+1}}{2n}, \qquad k = 1, \cdots, n. \tag{2}$$

If we take δ sufficiently small, we shall have

$$T_{n-1}(y_k) = \left(\frac{1}{2n} + \mu_k\right)(-1)^{k+1}, \qquad |\mu_k| < \epsilon, \qquad k = 1, \cdots, n. \tag{3}$$

We construct an even function $g \in C^*$, defining it on $[0, \pi]$ in the following way: We put $g(y_k) = (-1)^{k+1}, \ k = 1, \cdots, n, \ g(y) = 0$ on each of the intervals $[0, y_1 - \delta], \ [y_1 + \delta, y_2 - \delta], \ \cdots, \ [y_{n-1} + \delta, y_n - \delta]$; and we make $g(y)$ linear on the remaining intervals. It is easy to understand that $\omega(g, \pi/n) = 1$. On the other hand, by (3) and the definition of g,

$$g(0) - T_{n-1}(0) = -\frac{2n - 1}{2n},$$

$$g(y_k) - T_{n-1}(y_k) = \left(\frac{2n - 1}{2n} - \mu_k\right)(-1)^{k+1}, \ k = 1, \cdots, n.$$

Taking into account that g and T_{n-1} are even, we see that there are $2n$ points of K where $g(x) - T_{n-1}(x)$ takes values of alternating sign and of absolute value $\geqslant (2n - 1)/2n - \epsilon$. By Theorem 10, Chapter 2, $E^*_{n-1}(g) \geqslant (2n - 1)/(2n) - \epsilon$, and this is equivalent with (1). ∎

For an arbitrary ω, we are not able to find the smallest constant C for which the relation $E^*_{n-1}(\Lambda_\omega) \leqslant C\omega(\pi/n)$ is true. Nevertheless, we can find the best C that serves all ω at once.

THEOREM 10 (Korneǐčuk [69]). For all $f \in C^*$,

$$E^*_{n-1}(f) \leqslant \omega\left(f, \frac{\pi}{n}\right), \qquad n = 1, 2, \cdots. \tag{4}$$

If $M < 1$, then $E^*_{n-1}(f) \leqslant M\omega(f, \pi/n)$ cannot be true for all $f \in C^*$ and all $n = 1, 2, \cdots$.

Proof. The proof follows from Theorem 9 and the fact (Theorem 8, Chapter 3), that for each modulus of continuity ω there is a concave modulus of continuity ω_1 with $\omega(h) \leqslant \omega_1(h) \leqslant 2\omega(h)$. The last statement of the theorem follows from Lemma 3. ∎

Other than W^*_p (see Theorem 6), the class Λ_ω has no optimal polynomial operators.

THEOREM 11. Let ω be a concave modulus of continuity that is not linear on the interval $[0, \delta]$, $\delta = \pi/n$, for some $n = 1, 2, \cdots$. Then, for each linear polynomial operator L_{n-1} of degree $n - 1$,

$$E(\Lambda_\omega; L_{n-1}) > E^*_{n-1}(\Lambda_\omega). \tag{5}$$

Proof. Because ω is not linear on $[0, \delta]$,

$$\frac{\omega(t)}{t} > \frac{\omega(\delta)}{\delta}, \qquad 0 < t < \delta. \tag{6}$$

(Compare p. 44.) Assume that we have equality in (5). Let F denote the class of all functions from C^* that coincide with the function f_n of 5(6) outside $[-\frac{1}{2}\delta, \frac{1}{2}\delta]$, and which on this interval are increasing and have modulus of continuity not exceeding ω. Clearly, $F \subset \Lambda_\omega$ and, by Chebyshev's theorem, zero is the polynomial of best approximation of degree $n - 1$ of each function $f \in F$. On the other hand,

$$\|L_{n-1}(f) - f\| \leqslant E(\Lambda_\omega; L_{n-1}) = E^*_{n-1}(\Lambda_\omega) = \|f\|,$$

and since the polynomial of best approximation is unique,

$$L_{n-1}(f) = 0, \qquad f \in F. \tag{7}$$

In particular, let g be the linear interpolation of f_n on $[-\frac{1}{2}\delta, \frac{1}{2}\delta]$, and let h be linear between the points $-\frac{1}{2}\delta$, 0, and $\frac{1}{2}\delta$, with values

$$h(-\tfrac{1}{2}\delta) = f_n(-\tfrac{1}{2}\delta) = -\tfrac{1}{2}\omega(\delta), \qquad h(\tfrac{1}{2}\delta) = f_n(\tfrac{1}{2}\delta) = \tfrac{1}{2}\omega(\delta),$$

$$h(0) = \omega(\tfrac{1}{2}\delta) - \tfrac{1}{2}\omega(\delta) > 0$$

[see (6)]. Outside $[-\frac{1}{2}\delta, \frac{1}{2}\delta]$, $g(x) = h(x) = f_n(x)$. Then $g, h \in F$. It is sufficient to show that the moduli of continuity of the functions g, h on $[-\frac{1}{2}\delta, \frac{1}{2}\delta]$ do not exceed ω. For example,

$$\omega(h, t) = \frac{2\omega(\tfrac{1}{2}\delta)}{\delta} t \leqslant \omega(t), \qquad 0 \leqslant t \leqslant \tfrac{1}{2}\delta,$$

$$\omega(h, t) = \omega\left(\tfrac{1}{2}\delta\right) + \frac{\omega(\delta) - \omega(\tfrac{1}{2}\delta)}{\tfrac{1}{2}\delta}\left(t - \tfrac{1}{2}\delta\right) \leqslant \omega(t), \qquad \tfrac{1}{2}\delta \leqslant t \leqslant \delta,$$

by the concavity of ω.

According to (7), $L_{n-1}(f) = 0$, where $f = M(h - g)$, $M > 0$, is a function, linear on $[-\frac{1}{2}\delta, 0]$ and on $[0, \frac{1}{2}\delta]$, which vanishes outside $[-\frac{1}{2}\delta, \frac{1}{2}\delta]$ and satisfies $f(0) = M[\omega(\tfrac{1}{2}\delta) - \tfrac{1}{2}\omega(\delta)] > 0$. Adjusting M, we can assume that $f(0) = \omega(\tfrac{1}{2}\delta)$; then $f \in \Lambda_\omega$. We now have

$$\| L_{n-1}(f) - f \| = \| f \| = \omega(\tfrac{1}{2}\delta) > \tfrac{1}{2}\omega(\delta) = E^*_{n-1}(\Lambda_\omega),$$

by Theorem 9. This is a contradiction with our hypothesis.   ▌

7. Analytic Functions

Let $p = 0, 1, \cdots$ be an integer, let $\rho > 1$. By $A_p(\rho)$ we denote the set of all functions $f(z)$, analytic for $|z| < \rho$, which satisfy $|f^{(p)}(z)| \leqslant 1$, $|z| < \rho$. We put

$$\| f \| = \max_{|z| \leqslant 1} | f(z) | \tag{1}$$

and denote by $E_n(f)$ the degree of approximation of f by polynomials of degree n in the norm (1).

THEOREM 12 (Babenko [32]). For the class $A_p(\rho)$ one has

$$E_{n-1}(A_p(\rho)) = \frac{\rho^{p-n}}{n(n-1)\cdots(n-p+1)}, \qquad n \geqslant p. \tag{2}$$

An extremal function $f \in A_p(\rho)$, for which $E_{n-1}(f) = E_{n-1}(A_p(\rho))$, is $f(z) = Cz^n$, with an appropriate constant C.

We shall present here a simplification of Babenko's proof, due to Scheick, which at the same time gives more (Theorem 13 below).

Each function $f(z)$, analytic for $|z| < \rho$, $\rho > 1$, has a representation $f(z) = \sum_{k=0}^{\infty} c_k z^k$, where $c_k = f^{(k)}(0)/k!$; for $k \geqslant p$, we also have $c_k = a_k b_k$, where

$$a_k = \frac{1}{k(k-1)\cdots(k-p+1)}, \qquad k \geqslant p, \tag{3}$$

the coefficients b_k being given by

$$h(z) = z^p f^{(p)}(z) = \sum_{k=p}^{\infty} b_k z^k.$$

With these a_k, b_k,

$$f(z) = \sum_{k=0}^{n-1} c_k z^k + \sum_{k=n}^{\infty} a_k b_k z^k, \qquad n \geqslant p, \qquad |z| < \rho. \tag{4}$$

It will be convenient to consider a more general transformation of functions h. Let $g(z) = \sum_{k=0}^{\infty} a_k z^k$ be a function which is analytic for $|z| < 1$ and for which

$$G(r, \theta) = \tfrac{1}{2} a_n + \sum_{k=1}^{\infty} a_{n+k} r^k \cos k\theta \geqslant 0 \qquad \text{for } 0 \leqslant r < 1 \text{ and all real } \theta. \tag{5}$$

If $M > 0$, we denote by $B_g(\rho, M)$ the class of all functions $f(z)$, $|z| < \rho$, representable in the form

$$f(z) = \sum_{k=n}^{\infty} a_k b_k z^k, \qquad |z| < \rho, \tag{6}$$

where $h(z) = \sum_{0}^{\infty} b_k z^k$ is an arbitrary function in $|z| < \rho$, which satisfies $|h(z)| \leqslant M$, $|z| < \rho$. The series (6) converges, because for each z, $|z| < \rho$, we have $b_k z^k = O(q^k)$ with some q, $0 < q < 1$. We wish to find the degree of approximation of the class $B_g(\rho, M)$ by polynomials.

THEOREM 13 (Scheick [87]).   If the function g satisfies (5), then

$$E_{n-1}(B_g) = M a_n \rho^{-n}, \qquad n = 0, 1, \cdots; \tag{7}$$

an extremal function is $f(z) = M a_n \rho^{-n} z^n$.

LEMMA 4.   Let

$$G(\theta) = \tfrac{1}{2} a_n + \sum_{k=1}^{\infty} a_{n+k} \cos k\theta \geqslant 0 \quad \text{for all } \theta, \qquad \sum |a_k| < +\infty.$$

We define

$$\Phi(\theta) = \sum_{k=-\infty}^{n-1} \xi_k e^{-ik\theta} + \sum_{k=n}^{\infty} a_k e^{-ik\theta}, \qquad \sum |\xi_k| < +\infty. \tag{8}$$

Then the minimum of the integral $\int_{-\pi}^{\pi} |\Phi|\, d\theta$ for all ξ_k for which $\sum |\xi_k| < +\infty$ is equal to $2\pi a_n$; the minimum is attained for

$$\xi_k = a_{2n-k}, \qquad k \leqslant n-1. \tag{9}$$

Proof. On one hand,

$$\int_{-\pi}^{\pi} |\Phi|\, d\theta \geqslant \left| \int_{-\pi}^{\pi} e^{in\theta}\Phi(\theta)\, d\theta \right| = 2 \int_{-\pi}^{\pi} G(\theta)\, d\theta = 2\pi a_n .$$

On the other hand, with the choice (9),

$$\Phi(\theta) = \sum_{k=-\infty}^{n-1} a_{2n-k} e^{-ik\theta} + \sum_{k=n}^{\infty} a_k e^{-ik\theta} = 2e^{-in\theta}G(\theta),$$

and therefore in this case

$$\int_{-\pi}^{\pi} |\Phi|\, d\theta = 2 \int_{-\pi}^{\pi} G\, d\theta = 2\pi a_n . \qquad\qquad |$$

Proof of Theorem 13. Let $f \in B_g$, $1 < \rho_1 < \rho$, and let $|z| = 1$. We have

$$\int_{|t|=\rho_1} h(t)\, \bar{t}^k \frac{dt}{t} = \rho_1^{2k} \int_{|t|=\rho_1} h(t)\, t^{-k-1}dt = 0, \qquad k = -1, -2, \cdots . \tag{10}$$

Let ξ_k, $-\infty < k \leqslant n-1$ be arbitrary numbers for which $\sum |\xi_k| < +\infty$. By (6) and (10), after the substitution $t = z\rho_1 e^{i\theta}$, we have

$$f(z) + \sum_{k=0}^{n-1} b_k \xi_k \rho_1^k z^k$$

$$= \frac{1}{2\pi i} \int_{|t|=\rho_1} h(t) \left\{ \sum_{k=-\infty}^{-1} \xi_k \rho_1^{-k} \overline{\left(\frac{z}{t}\right)}^{-k} + \sum_{k=0}^{n-1} \xi_k \rho_1^k \left(\frac{z}{t}\right)^k + \sum_{k=n}^{\infty} a_k \left(\frac{z}{t}\right)^k \right\} \frac{dt}{t}$$

$$= \frac{1}{2\pi} \int_{-\pi}^{\pi} h(z\rho_1 e^{i\theta}) \left\{ \sum_{k=-\infty}^{n-1} \xi_k e^{-ik\theta} + \sum_{k=n}^{\infty} a_k \rho_1^{-k} e^{-ik\theta} \right\} d\theta. \tag{11}$$

If $\Phi(\theta)$ is the expression in the last curved bracket, then

$$\left| f(z) + \sum_{k=0}^{n-1} b_k \xi_k \rho_1^k z^k \right| \leqslant \frac{M}{2\pi} \int_{-\pi}^{\pi} |\Phi(\theta)|\, d\theta.$$

In particular, we select

$$\xi_k = a_{2n-k}\rho_1^{2n-k}, \qquad -\infty < k \leqslant n-1. \tag{12}$$

Then by Lemma 4, for our fixed ρ_1,

$$\left| f(z) + \sum_{k=0}^{n-1} b_k a_{2n-k} \rho_1^{-2(n-k)} z^k \right| \leqslant M a_n \rho_1^{-n}, \qquad |z| = 1. \tag{13}$$

We can now make $\rho_1 \to \rho$ and find

$$|f(z) - P_{n-1}(z)| \leqslant M a_n \rho^{-n}, \qquad |z| = 1, \tag{14}$$

where $P_{n-1}(z)$ is the polynomial $-\sum_{0}^{n-1} b_k a_{2n-k} \rho^{-2(n-k)} z^k$. By the maximum modulus principle, the inequality (14) holds also for $|z| \leqslant 1$. Thus

$$E_{n-1}(f) \leqslant M a_n \rho^{-n}, \qquad f \in B_g.$$

It remains to remark that the function $f(z) = M a_n \rho^{-n} z^n$ (which corresponds to $h(z) = M \rho^{-n} z^n$ in (6)) belongs to B_g and that its polynomial of best approximation of degree $n - 1$ is zero (see p. 32). ▌

Proof of Theorem 12. We start by giving sufficient conditions for a function

$$G(x) = \tfrac{1}{2} a_0 + \sum_{k=1}^{\infty} a_k \cos kx$$

to be positive for all x. This is true if the coefficients a_k are positive, convex, that is, satisfy $\Delta^2 a_k = a_k - 2a_{k+1} + a_{k+2} \geqslant 0$, $k = 0, 1, \cdots$, and if $\sum a_k < +\infty$. The proof consists in a double partial summation. Using the kernels D_n [1.2(10)] and

$$\tilde{D}_n(x) = D_0(x) + \cdots + D_n(x) = \frac{1 - \cos nx}{4 \sin^2 (x/2)} \geqslant 0$$

[compare 1.2(11)] we find (with $\Delta a_k = a_k - a_{k+1}$) that

$$\tfrac{1}{2} a_0 + \sum_{k=1}^{n} a_k \cos kx$$

$$= \sum_{k=0}^{n-1} \Delta a_k D_k(x) + a_n D_n(x)$$

$$= \sum_{k=0}^{n-2} \Delta^2 a_k \tilde{D}_k(x) + \Delta a_{n-1} \tilde{D}_{n-1}(x) + a_n D_n(x), \qquad 0 < x < 2\pi.$$

The last two terms tend to zero when $n \to \infty$, so that

$$G(x) = \sum_{k=0}^{\infty} \Delta^2 a_k \tilde{D}_k(x) \geqslant 0$$

for all x.

Assume now that a_n is a sequence which satisfies

$$a_n \geqslant 0, \qquad \Delta a_n \geqslant 0, \qquad \Delta^2 a_n \geqslant 0, \qquad n = 0, 1, \cdots. \tag{15}$$

In this case

$$\tfrac{1}{2} a_0 + \sum_{k=1}^{\infty} a_k r^k \cos kx \geqslant 0 \qquad \text{for } 0 \leqslant r < 1, \text{ and all } x.$$

For

$$\Delta^2(r^n a_n) = r^n \Delta^2 a_n + 2r^n(1 - r) \Delta a_{n+1} + r^n(1 - r)^2 a_{n+2} \geqslant 0.$$

Let $f \in A_p(\rho)$. From (3) it follows that for some polynomial P_{n-1}, $f - P_{n-1} \in B_g(\rho, \rho^p)$, where $g(z)$ is the series $\sum_{k=n}^{\infty} a_k z^k$ with the coefficients (3). The coefficients a_n satisfy (15), therefore we obtain from (7) that

$$E_{n-1}(f) \leqslant \frac{\rho^{p-n}}{n(n - 1) \cdots (n - p + 1)}, \qquad n \geqslant p. \tag{16}$$

It remains to remark that the function $f(z) = \rho^{p-n} z^n / [n \cdots (n - p + 1)]$ belongs to $A_p(\rho)$ and that its degree of approximation $E_{n-1}(f)$ is equal to the right-hand side of (16). |

8. Notes

1. From Theorem 1 it is easy to derive [15, p. 38] the following important theorem of D. Jackson:

For a Chebyshev system $\phi_1, \cdots, \phi_n$, defined on $[a, b]$ or K, the polynomial of best approximation in the L^1 norm is unique for each continuous function.

2. Korneĭčuk [67, 70] has obtained the expressions for $E_{n-1}^*(\Lambda_{p\omega})$ in the case when $p = 1, 2, 3$ and ω is a concave modulus of continuity. In all three cases, $E_{n-1}^*(\Lambda_{p\omega}) = \| f_{np} \|$, where f_n stands for the function 5(6), and g_p denotes the pth integral on K, with mean-value zero, of a function g. It is not known whether this holds also for $p > 3$.

PROBLEMS

1. Find an example of a polynomial P_n, of a function $f \in C[a, b]$ and of a Chebyshev system $\phi_1, \cdots, \phi_n$ with the property that

$$\int_a^b \phi_k \operatorname{sign} (f - P_n) \, dx = 0, \qquad k = 1, \cdots, n,$$

although P_n is not a polynomial of best L^1 approximation.

2. Prove that actually there is an equality sign in 4(6).

3. Let ω be a concave modulus of continuity, let f be a function that belongs to Λ_ω on $A = [a, b]$ or $A = K$. If $(\alpha, \beta) \subset A$ and if f_1 is the continuous function that coincides with f outside (α, β) and is linear on (α, β), then $f_1 \in \Lambda_\omega$.

4. If L_n is a linear trigonometric operator of degree n, which leaves invariant all T_n, then for a concave nonlinear ω,

$$E(\Lambda_\omega, L_{n-1}) \geqslant \frac{n}{\pi} \int_0^{\pi/n} \omega(t) \, dt > E_{n-1}^*(\Lambda_\omega).$$

5. Formula 7(2) is true also for nonintegral $p > 0$, if one defines the pth derivative of the function $f(z) = \sum_0^\infty c_k z^k$ by

$$z^p f^{(p)}(z) = \sum_{k=0}^\infty \frac{\Gamma(k+1)}{\Gamma(k-p+1)} c_k z^k.$$

6. Let T be the differential operator defined by $T(f, z) = zf'(z)$. It maps into itself the space H of functions $f(z)$ analytic in the disc $|z| < 1$. If $G(x)$ is defined for $x \geqslant n$ and satisfies $\lim_{k \to \infty} |G(k)| = 1$, we put

$$G(T)(f, z) = \sum_{k=n}^\infty G(k) c_k z^k, \qquad f(z) = \sum_0^\infty c_k z^k \in H.$$

If in addition

$$G(x) > 0, \qquad G'(x) > 0, \qquad 2G'(x)^2 - G(x)G''(x) > 0, \qquad x > n,$$

then $E_{n-1}(S_G, r) = r^n/G(n)$, $0 < r \leqslant 1$, where S_G is the class of all functions $f \in H$ whose $G(T)$ transform satisfies $|G(T)(f, z)| \leqslant 1$, $|z| < 1$. (Scheick).

Widths

1. Definitions and Basic Properties

From the abstract point of view, approximation by polynomials or by trigonometric polynomials is a very special process. It is natural to try approximation by other systems of functions. For a given class of functions A, we can even try to find a "most favorable" system of approximation. One must bear in mind, however, that if A is a single function, the latter problem does not arise: The degree of approximation of the function f is zero if f itself is included in the system.

The following definitions will make our ideas more precise. Let X be a Banach space, and A and A' two subsets of X. The deviation of A from A' [see 8.1(2)] is the number

$$E_{A'}(A) = \sup_{f \in A} \{ \inf_{g \in A'} \| f - g \| \}. \tag{1}$$

In particular, let $A' = X_n$ be the n-dimensional subspace of X spanned by the elements $\phi_1, \cdots, \phi_n$. $E_{X_n}(A)$ is the degree of approximation of the class A by linear combinations $a_1 \phi_1 + \cdots + a_n \phi_n$. It was Kolmogorov's idea [60] to consider the infimum of this deviation for all n-dimensional subspaces X_n of X:

$$d_n(A) = \inf_{X_n} E_{X_n}(A), \qquad n = 0, 1, 2, \cdots. \tag{2}$$

According to (2), $d_n(A) = 0$ for each A if X is of dimension n. If n exceeds the dimension of X, we define $d_n(A) = 0$. The number $d_n(A) = d_n^X(A)$ is called the nth *width of A in X*. The definition (2) is sensible if the origin in X is in some sense the center of A; otherwise, one should replace in the definition (2) the subspaces X_n by all possible translations of n-dimensional subspaces. For example, $d_0(A)$ is the radius of the smallest closed ball, with center at the origin, that contains A. If all translations of the subspaces X_0 are allowed, then $d_0(A)$ becomes the radius of the smallest closed ball of arbitrary center that contains A.

Obviously, we have $d_n(A) \leqslant d_n(A')$ if $A \subset A'$. Also,

$$d_0(A) \geqslant d_1(A) \geqslant d_2(A) \geqslant \cdots. \tag{3}$$

If *A is compact*, then $d_n(A) \to 0$. Indeed, for each $\epsilon > 0$, A has a finite ϵ-net.

If this net consists of n elements, then the subspace Y that they span has dimension $k \leqslant n$, and $d_n(A) \leqslant E_Y(A) \leqslant \epsilon$. If U is the closed unit ball in X, with center in the origin, and if X is more than n-dimensional, then

$$d_n(U) = 1. \tag{4}$$

This can be proved in the following way: Let X_n be some n-dimensional subspace of X, and let y be a point of X that does not belong to X_n. There exists (Theorem 1, Chapter 2) a point $x \in X_n$ with

$$\| y - x \| = \rho(y, X_n) = a > 0.$$

Put $z = (y - x)/a$. Obviously, $\| z \| = 1$, and $\rho(z, X_n) = 1$. Hence, $E_{X_n}(U) = 1$ for each subspace X_n.

If the infimum in (2) is attained for some X_n, this subspace is called an *extremal subspace*. Kolmogorov insists that the determination of the *exact* value of $d_n(A)$ is important because it may lead to the discovery of extremal subspaces, and therefore to new and better methods of approximation.

In this chapter we shall discuss several methods that allow an exact or an asymptotic determination of the $d_n(A)$. We begin with the simplest approach.

2. Sets of Continuous and Differentiable Functions

In order to estimate the widths of a class $A \subset C[B]$, one can proceed in the following way: Upper bounds for the $d_n(A)$ can often be derived from classical results. Often we know the degree of approximation $E_{X_n}(A)$ of A for some concrete spaces X_n; for instance, for spaces of polynomials, or trigonometric polynomials. This gives an upper bound for the widths, $d_n(A) \leqslant E_{X_n}(A), n = 0, 1, \cdots$. A simple way to obtain lower bounds is

LEMMA 1. Let B be a compact Hausdorff space. Assume that there exist $n + 1$ points x_i, $i = 0, \cdots, n$ of B and a number $\epsilon \geqslant 0$ with the following property: For each distribution of signs $\lambda_i = \pm 1, i = 0, \cdots, n$, there is a function $f_0 \in A$ such that

$$\operatorname{sign} f_0(x_i) = \lambda_i, \qquad | f_0(x_i) | \geqslant \epsilon, \qquad i = 0, \cdots, n. \tag{1}$$

Then, in the space $C[B]$,

$$d_n(A) \geqslant \epsilon. \tag{2}$$

Proof. Let X_n be a subspace of C spanned by $\phi_1, \cdots, \phi_n$. The system of n linear equations in $n + 1$ unknowns,

$$\sum_{i=0}^{n} c_i \phi_k(x_i) = 0, \qquad k = 1, 2, \cdots, n,$$

has a nontrivial solution $c_0, c_1, \cdots, c_n$ such that $\sum_0^n |c_i| = 1$. With λ_i chosen so that $\lambda_i c_i \geqslant 0$, $i = 0, \cdots, n$, we take $f_0 \in A$ satisfying (1). Then, regardless of how the a_k, $k = 1, 2, \cdots, n$, are taken, we have

$$\left\| f_0 - \sum_{k=1}^n a_k \phi_k \right\| \geqslant \sum_{i=0}^n |c_i| \left| f_0(x_i) - \sum_{k=1}^n a_k \phi_k(x_i) \right|$$

$$\geqslant \left| \sum_{i=0}^n c_i f_0(x_i) - \sum_{k=1}^n a_k \sum_{i=0}^n c_i \phi_k(x_i) \right|$$

$$\geqslant \epsilon \sum_{i=0}^n |c_i| = \epsilon. \qquad \blacksquare$$

In particular, we shall apply Lemma 1 to the classes $\Lambda_{p\omega}^s = \Lambda_{p\omega}^s(M_0, \cdots, M_{p+1}; B)$ of functions $f(x) = f(x_1, \cdots, x_s)$ on the s-dimensional parallelepiped B or the s-dimensional torus (see Chapter 3, Sec. 7). The following lemma deals with functions of one variable.

LEMMA 2.   Let a positive integer p and a modulus of continuity $\omega(h)$, $0 \leqslant h \leqslant l$ be given. There exist constants $C_p > 0$, $C_p' > 0$ with the following properties: For each $0 < \delta \leqslant l$, there is a function $g(x) = g_p(x)$, $-\infty < x < +\infty$, with p continuous derivatives, that vanishes outside $(-\delta, \delta)$, increases on $(-\delta, 0)$, decreases on $(0, \delta)$, and satisfies:

$$\| g_p^{(k)} \| \leqslant C_p \delta^{p-k} \omega(\delta), \qquad k = 0, \cdots, p; \tag{3}$$

$$\omega(g_p^{(k)}, t) \leqslant C_p \delta^{p-k} \omega(t); \qquad k = 0, \cdots, p; \tag{4}$$

$$\| g_p \| = g_p(0) \geqslant C_p' \delta^p \omega(\delta). \tag{5}$$

Proof.   We prove even more than (5), namely,

$$g_p(-\tfrac{1}{2}\delta) \geqslant C_p' \delta^p \omega(\delta). \tag{6}$$

The function g_p and the constants C_p, C_p' will be defined by induction in p. For $p = 0$, we can take $g_0(x) = \omega(x + \delta)$, $-\delta \leqslant x \leqslant 0$; $g_0(x) = \omega(\delta - x)$, $0 \leqslant x \leqslant \delta$; $g_0(x) = 0$ elsewhere. Then (3), (4), and (6) hold with $C_0 = 1$, $C_0' = \tfrac{1}{2}$ [for (6), we use 3.5(3)].
   If $p > 0$, and g_{p-1} is already known, we define

$$h(x) = \begin{cases} g_{p-1}(2x + \delta), & -\delta \leqslant x \leqslant 0; \\ -g_{p-1}(2x - \delta), & 0 \leqslant x \leqslant \delta; \\ 0 & \text{elsewhere.} \end{cases}$$

Since $g_{p-1}(x)$ has $p-1$ continuous derivatives, which all vanish for $x = \pm \delta$, h has $p-1$ continuous derivatives on $(-\infty, +\infty)$. We put

$$g_p(x) = \int_{-\delta}^{x} h(t) \, dt. \tag{7}$$

It follows that g_p has p continuous derivatives that vanish outside $(-\delta, +\delta)$. From (7) we derive

$$g_p(-\tfrac{1}{2}\delta) \geqslant \int_{-(3/4)\delta}^{-(1/2)\delta} h(t) \, dt \geqslant \tfrac{1}{4}\delta h(-\tfrac{3}{4}\delta) = \tfrac{1}{4}\delta g_{p-1}(-\tfrac{1}{2}\delta) \geqslant \tfrac{1}{4} C'_{p-1}\delta^p \omega(\delta),$$

so that (6) holds, with $C'_p = \tfrac{1}{4} C'_{p-1}$.

A similar computation shows that $g_p(0) \leqslant \delta g_{p-1}(0)$, so that

$$g_p(0) = \| g_p \| \leqslant \delta^p \omega(\delta).$$

Differentiating (7), we obtain $g_p^{(k)}(x) = \pm 2^{k-1} g_{p-1}^{(k-1)}(2x \pm \delta)$ or $= 0$, so that $g_p^{(k)}(x)$ is of the form $\pm c g_{p-k}(2^k(x + b))$, or $= 0$. Hence, we have (3), if C_p is large enough.

For $k = p$, we see that $g_p^{(p)}(x)$ is zero except for a finite number of intervals I_i, on each of which it has the form $\pm c g_0(2^p(x + b_i))$. Recalling the definition of g_0, we see that $g_p^{(p)}$ is one of the functions 3.5(7), with $\omega(x)$ replaced by $\omega(2^p x)$. Using 3.5(8), $\omega(g_p^{(p)}, h) \leqslant c \omega(2^p h) \leqslant 2^p c \omega(h)$.

We have proved (4) for $k = p$. If $k < p$, and if (3) holds with some C_p, we prove that $\omega(g_p^{(k)}, t) \leqslant 3 C_p \delta^{p-k} \omega(t)$. Since $g_p^{(k)}$ is zero outside $(-\delta, \delta)$, it is sufficient to show this only for $0 \leqslant t \leqslant 2\delta$. We have, by 3.5(4),

$$| g_p^{(k)}(x + t) - g_p^{(k)}(x) | \leqslant t \| g_p^{(k+1)} \| \leqslant C_p t \delta^{p-k-1} \omega(\delta)$$

$$\leqslant C_p \delta^{p-k} \left(1 + \frac{t}{\delta} \right) \omega(t) \leqslant 3 C_p \delta^{p-k} \omega(t). \qquad \blacksquare$$

THEOREM 1.   The widths of the classes $\Lambda_{p\omega}^s$ are

$$d_n(\Lambda_{p\omega}^s) \approx n^{-p/s} \omega(n^{-1/s}). \tag{8}$$

Proof.   The estimate of $d_n(\Lambda)$ from above follows from Theorems 7 and 8, Chapter 6.

Let l denote the smallest of the side lengths of B, $l = \min (b_k - a_k)$ if B is a parallelopiped, $l = 2\pi$ if $B = K^{(s)}$. For $n = 1, 2, \cdots$, each side contains $\geqslant n^{1/s}$ disjoint open intervals of length 4δ, where

$$\delta = \tfrac{1}{4} l([n^{1/s}] + 1)^{-1} \geqslant C n^{-1/s} \tag{9}$$

and $C > 0$ is a constant. Therefore B contains $n = (n^{1/s})^s$ disjoint cubes, with sides parallel to coordinate axes of side length 4δ. Let $z_1, \cdots, z_n$ be their centers, and U_i the balls of centers z_i and radii δ. Let $\epsilon = \delta^p \omega(\delta)$, and

$$G(x) = \epsilon^{1-s} \prod_{i=1}^{s} g_p(x_i), \qquad x = (x_1, \cdots, x_s), \tag{10}$$

with g_p given by Lemma 2.

The function G has continuous partial derivatives of orders $k \leqslant p$:

$$D^k G(x) = \epsilon^{1-s} \prod_{i=1}^{s} g_p^{(k_i)}(x_i), \qquad \sum_{1}^{s} k_i = k. \tag{11}$$

Applying (3) of Lemma 2, to each factor of the product, we obtain

$$\| D^k G \| \leqslant M, \qquad k \leqslant p \tag{12}$$

if $M \geqslant C_p^s l^{p-k} \omega(l)$. Thus, we can find an M that depends upon p, s, and ω, but not on n or δ.

If $k = p$ in (11), then $\sum k_i = p$, and we obtain [by 3.6(5), (3), and (4)],

$$\omega(D^p G, h) \leqslant \epsilon^{1-s} \sum_{i=1}^{s} C_p^s \delta^{sp-p} \omega(\delta)^{s-1} \omega(h) = s C_p^s \omega(h) \leqslant M \omega(h), \tag{13}$$

if $M \geqslant s C_p^s$. Finally, from (5), we derive

$$G(0) \geqslant M' \delta^p \omega(\delta) = M' \epsilon \tag{14}$$

with $M' = C_p'^s$. Inequalities (12) to (14) show that, for sufficiently small $\alpha > 0$, and $m > 0$, the function $H(x) = \alpha G(x)$ has the properties:

$$\| D^k H \| \leqslant M_k, \qquad k = 0, \cdots, p;$$

$$\omega(D^p H, h) \leqslant \tfrac{1}{2} \omega(h), \tag{15}$$

$$H(0) \geqslant m\epsilon.$$

We consider the functions

$$f_0(x) = \sum_{i=1}^{n} \lambda_i H(x - z_i), \qquad \lambda_i = \pm 1. \tag{16}$$

It is easy to see [compare 3.6(9)] that $\omega(f_0, h) \leqslant 2\omega(H, h) \leqslant M_{p+1} \omega(h)$. This and the first relation (15) show that all functions f_0 belong to the class $\Lambda_{p\omega}^s$. We can apply Lemma 1 and obtain $d_n(\Lambda) \geqslant m\epsilon \geqslant \text{const } n^{-p/s} \omega(n^{-1/s})$. ∎

3. Widths of Balls

The following theorem is often useful.

THEOREM 2 (Gohberg and Kreĭn [53]). Let X_{n+1} be an $(n + 1)$-dimensional subspace of a Banach space X, and let U_{n+1} be the closed unit ball of X_{n+1}. Then

$$d_n^X(U_{n+1}) = 1. \tag{1}$$

In the case when $X = X_{n+1}$, this follows from 1(4), and has a simple proof. The general case is deeper. Its proof will be based on a topological theorem of Borsuk.

Let $\Sigma_n: x_1^2 + \cdots + x_{n+1}^2 = 1$ denote the unit sphere of the $(n + 1)$-dimensional euclidean space E_{n+1}. We shall consider n-dimensional vector fields on Σ_n. Each such vector field is given by a function $P(y)$, $y \in \Sigma_n$, whose values are vectors, which belong to a fixed n-dimensional linear space X_n. One can make X_n a Banach space by introducing a norm in X_n, in many different ways. For example, one can put $\| x \| = \sqrt{x_1^2 + \cdots + x_n^2}$, where x_k are the coordinates of $x \in X_n$ with respect to some basis. However, by a useful theorem (which can be derived from Lemma 1, Chapter 2) all these Banach spaces are homeomorphic to each other. We select one of these norms for X_n.

THEOREM OF BORSUK. If $P(y)$ is a continuous n-dimensional vector field on Σ_n, and if $P(y)$ is odd, that is, $P(-y) = -P(y)$, for all $y \in \Sigma_n$, then $P(y)$ vanishes at some point of Σ_n.

We cannot give a proof of this theorem here. The reader is referred to Tucker [98] and Whittlesey [104, p. 815].

Proof of Theorem 2. We first note that the theorem of Borsuk remains true if Σ_n is replaced by the sphere $S_n: \| y \| = 1$ of an $(n + 1)$-dimensional Banach space X_{n+1}.

To show this, we consider an isomorphic mapping of X_{n+1} onto E_{n+1}. It maps S_n onto a subset S_n' of E_{n+1}, transforming the function P on S_n into P' on S_n', which is again odd and continuous. Now the set S_n' has exactly one point $(x_1, \cdots, x_{n+1})$ on each ray through the origin. We map S_n' continuously onto Σ_n by means of the correspondence

$$(x_1, \cdots, x_{n+1}) \rightarrow \frac{(x_1, \cdots, x_{n+1})}{\sqrt{x_1^2 + \cdots + x_{n+1}^2}}.$$

Then P' becomes an odd, continuous, n-dimensional vector field π on Σ_n. Applying Borsuk's theorem to π, we see that it holds also for P on S_n.

Since $d_n(U_{n+1}) \leqslant d_0(U_{n+1}) = 1$, our statement (1) will follow if we can show that for each n-dimensional subspace X_n of X, there is an element y for which

$$y \in X_{n+1}, \qquad \| y \| = 1, \qquad \rho(y, X_n) = 1. \tag{2}$$

Proving (2), we shall assume that X is finite dimensional. This is no restriction of generality, for we can replace X by its smallest subspace (of dimension at most $2n + 1$) that contains X_n and X_{n+1}. We consider two cases.

(a)   The norm of the space X is strictly convex, that is, the sphere $\| x \| = 1$ in X does not contain segments. This can be expressed in the following way:

$$\left. \begin{array}{c} x \neq y, \qquad \| x \| = \| y \| = 1, \qquad \alpha, \beta > 0, \qquad \alpha + \beta = 1 \\[2mm] \text{imply } \| \alpha x + \beta y \| < 1. \end{array} \right\} \tag{3}$$

In this case we know (Chapter 2, Sec. 1) that for each $y \in X$ there is a unique $x \in X_n$ of best approximation, so that

$$\| y - x \| = \rho(y, X_n).$$

We shall denote this x by $P(y)$.

We observe that the function $P(y)$ is *continuous*. Indeed, let $P(y) = x$ and $y_n \to y$. Assume that $P(y_n) \nrightarrow x$. Because X_n is locally compact, we can assume (replacing, if necessary, y_n by a subsequence) that $P(y_n) \to x' \neq x$. But then $\| y_n - x \| \to \| y - x \|$ and $\| y_n - P(y_n) \| \to \| y - x' \| > \| y - x \|$ imply that, for large n, $\| y_n - x \| < \| y_n - P(y_n) \|$, in contradiction to the definition of P.

Next, $P(y)$ is *odd*. Let $P(y_0) = x_0$; then

$$\| (- y_0) - (- x_0) \| = \| y_0 - x_0 \| = \inf_{x \in X_n} \| y_0 - x \|$$

$$= \inf_{x \in X_n} \| (- y_0) - x \| = \rho(- y_0, X_n),$$

so that $P(- y_0) = - x_0$.

We now apply Borsuk's theorem to the sphere S_n: $\| y \| = 1$ in X_{n+1}. There exists a $y \in S_n$ with $P(y) = 0$, and we obtain (6):

$$\rho(y, X_n) = \rho(y, P(y)) = \rho(y, 0) = \| y \| = 1.$$

(b)   If the norm of X is not strictly convex, we alter it slightly. The condition (3) of strict convexity is equivalent to the following property:

$$\left. \begin{array}{l} \text{If the vectors } x', y' \in X \text{ are different from zero, and one is not} \\ \text{a positive multiple of the other, then} \\[2mm] \qquad \| x' + y' \| < \| x' \| + \| y' \|. \end{array} \right\} \tag{4}$$

It is obvious that (4) implies (3) because, under the assumptions of (3), αx and

βy are not positive multiples of each other ($\alpha x = c\beta y$, $c \geqslant 0$ would imply $\alpha = c\beta$ and $x = y$). Also, (3) implies (4); this is seen by taking in (4),

$$x = \frac{x'}{\| x' \|}, \qquad\qquad y = \frac{y'}{\| y' \|}$$

and

$$\alpha = \frac{\| x' \|}{\| x' \| + \| y' \|}, \qquad \beta = \frac{\| y' \|}{\| x' \| + \| y' \|}.$$

We now approximate the norm of X closely with a strictly convex norm. Let $\epsilon > 0$ be given. Since X is finite dimensional, there is a strictly convex norm $\| x \|_1$ in X, for instance, the euclidean norm, taken with respect to an arbitrarily chosen basis. Since $\| x \|_1$ is continuous with respect to $\| x \|$, by the theorem mentioned on page 137, it is bounded, say, $\| x \|_1 \leqslant M$, on the compact set $\| x \| \leqslant 1$. Changing $\| x \|_1$ by a constant factor, we may assume that $\| x \|_1 \leqslant \epsilon \| x \|$, $x \in X$. We then put

$$\| x \|_0 = \| x \| + \| x \|_1. \tag{5}$$

The norm $\| x \|_0$ satisfies (4) and so is strictly convex. Also,

$$\| x \| \leqslant \| x \|_0 \leqslant (1 + \epsilon) \| x \|, \qquad x \in X.$$

Applying (a) to X with the norm $\| x \|_0$, we see that there is an element $y_\epsilon \in X_{n+1}$ for which $(1 + \epsilon)^{-1} \| y_\epsilon \| \leqslant 1$, $(1 + \epsilon)^{-1} \leqslant \rho(y_\epsilon, X_n) \leqslant 1$. Making $\epsilon \to 0$ and using the local compactness of X, we find an element y that satisfies (2). ∎

4. Applications of Theorem 2

Let X be a Banach space, and let $\Phi: \phi_1, \cdots, \phi_n, \cdots$, be a sequence of linearly independent elements of X. Let Δ be a sequence $\delta_0 \geqslant \cdots \geqslant \delta_n \geqslant \cdots$, for which $\delta_n > 0$, $\delta_n \to 0$ as $n \to \infty$. We allow some of the δ_n to be $+ \infty$. Let X_n denote the n-dimensional subspace spanned by $\phi_1, \cdots, \phi_n$.

We define the set $A = A(\Phi, \Delta)$ to consists of all elements $x \in X$ for which

$$E_{X_n}(x) \leqslant \delta_n, \qquad n = 0, 1, \cdots. \tag{1}$$

Here, $E_{X_n}(x) = \rho(x, X_n)$ is the degree of approximation of x by linear combinations of $\phi_1, \cdots, \phi_n$; in particular, $E_{X_0}(x) = \| x \|$. Sets A of this type are called *full approximation sets*.

THEOREM 3. For a full approximation set $A(\Phi, \Delta)$,

$$d_n(A) = \delta_n, \qquad n = 0, 1, \cdots. \tag{2}$$

The subspace X_n is extremal for each n.

Proof. We have $d_n(A) \leqslant E_{X_n}(A) \leqslant \delta_n$, by (1). On the other hand, let U be the ball $\| x \| \leqslant \delta_n$ of X_{n+1}. Then, for $x \in U$, $E_{X_k}(x) \leqslant \delta_n \leqslant \delta_k$, $k = 0, \cdots, n$; and $E_{X_k}(x) = 0$, $k = n + 1, \cdots$. Theorefore, $U \subset A$, and by Theorem 2,

$$d_n^X(A) \geqslant d_n^X(U) = \delta_n.$$

This establishes (2). Also,

$$E_{X_n}(A) = d_n(A), \qquad n = 0, 1, \cdots. \qquad \blacksquare$$

This proof has been based on the facts that (1) holds for each $x \in A$; and that the ball $\| x \| \leqslant \delta_n$ of X_{n+1} is contained in A. There are other important sets that satisfy these conditions.

The space l^2 consists of all points $x = \{a_k\}$, with real coordinates a_k, $k = 1, 2, \cdots$, for which

$$\| x \| = \sqrt{\sum_1^\infty a_k^2} < +\infty.$$

An *ellipsoid* $D = D(\varDelta)$ in l^2, where $\varDelta$ is a sequence $\delta_1 \geqslant \delta_2 \geqslant \cdots, 0 < \delta_k \leqslant +\infty$ $k = 1, 2, \cdots$, is determined by the inequality

$$\sum_{k=1}^\infty \delta_k^{-2} a_k^2 \leqslant 1. \tag{3}$$

THEOREM 4

$$d_n(D) = \delta_{n+1}, \qquad n = 0, 1, \cdots. \tag{4}$$

Proof. Let ϕ_n, $n = 1, 2, \cdots$, have the nth coordinate equal to 1 and all other coordinates equal to zero. If $x \in D$, then

$$\rho(x, X_n)^2 \leqslant \sum_{k=n+1}^\infty a_k^2 \leqslant \delta_{n+1}^2 \sum_{n+1}^\infty \delta_k^{-2} a_k^2 \leqslant \delta_{n+1}^2;$$

hence $E_{X_n}(x) \leqslant \delta_{n+1}$. If $x \in X_{n+1}$, then $a_{n+2} = \cdots = 0$, and if in addition $\| x \| \leqslant \delta_{n+1}$, then

$$\sum_{k=1}^\infty \delta_k^{-2} a_k^2 \leqslant \delta_{n+1}^{-2} \sum_{k=1}^\infty a_k^2 \leqslant 1,$$

so that $x \in D$. $\qquad \blacksquare$

We turn our attention to classes $\Re$ of functions, in particular to classes of periodic functions. Here, too, Theorem 2 often allows an exact determination of the widths. The knowledge of the extremal functions F of $\Re$, that is, functions that provide maxima for the degree of approximation, is very helpful.

Let $F(x)$ be a function bounded and measurable on K, but not necessarily continuous, that is similar to $\sin nx$ in the following sense: It is $(2\pi/n)$-periodic, odd, symmetric with respect to $x = \frac{1}{2}\pi/n$, positive and increasing on $(0, \frac{1}{2}\pi/n)$. It is clear that $F(x)$ has the sign $(-1)^{i-1}$ on the interval

$$I_i = \left(\frac{(i-1)\pi}{n}, \frac{i\pi}{n} \right), \qquad i = 1, \cdots, 2n;$$

$|F(x)|$ attains $2n$ maxima with values $F(\frac{1}{2}\pi/n)$ on K.

Since $\int_K F \, dx = 0$, F has an indefinite integral with mean-value zero (see Chapter 1, Sec. 1); it is given by

$$F_1(x) = \int_{\pi/(2n)}^{x} F(t) \, dt.$$

The function F_1 has a character similar to F, with a shift to the right by the amount $\frac{1}{2}\pi/n$: It is $(2\pi/n)$-periodic, odd with respect to $\frac{1}{2}\pi/n$, symmetric with respect to π/n; it is positive and increasing on $(\frac{1}{2}\pi/n, \pi/n)$. The pth indefinite integral with mean-value zero $F_p(x)$ behaves in the same way. In particular, F_p attains $2n$ extrema on K of the same absolute value and of alternating signs.

Let $g_i(x) = |F(x)|$, $x \in I_i$, $g_i(x) = 0$ for $x \notin I_i$, $i = 1, \cdots, 2n$. A function of the form $f = \sum\limits_{i=1}^{2n} a_i g_i$ has an indefinite integral f_1 on K if and only if

$$\sum_{i=1}^{2n} a_i = 0. \tag{5}$$

For a given integer $p = 0, 1, \cdots$, let X_{2n} denote the subspace of C^*, formed by all indefinite pth integrals f_p (with arbitrary mean values) of the functions f. (For $p = 0$, $f_p = f$.) Considering the cases $p = 0$, $p > 1$ separately, we see that X_{2n} is of dimension $2n$.

LEMMA 3.   Let p be a positive integer. If a class $\Re \subset C^*$ contains all $f_p \in X_{2n}$, which correspond to values a_i with $|a_i| \leqslant 1$, $i = 1, \cdots, 2n$, then

$$d_{2n-1}(\Re) \geqslant \|F_p\|. \tag{6}$$

Proof.   We begin by showing that

$$f_p \in X_{2n}, \qquad \|f_p\| \leqslant \|F_p\| \qquad \text{imply} \qquad |a_i| \leqslant 1, \qquad i = 1, \cdots, 2n. \tag{7}$$

We can assume $p > 0$. If (7) does not hold, there is a function $f_p \in X_{2n}$ with $\|f_p\| \leqslant \|F_p\|$ and $|a_k| > 1$ for some k. Multiplying this f_p with some constant q, $|q| < 1$, we obtain another $f_p \in X_{2n}$, an indefinite pth integral of a function $f = \sum\limits_{1}^{2n} a_i g_i$, for which $\|f_p\| < \|F_p\|$, $|a_i| \leqslant 1$, $i = 1, \cdots, 2n$, and $a_k = 1$ for some k.

Because of the properties of F_p and f_p, the difference $h(x) = F_p(x) - f_p(x)$ changes sign at least $2n$ times on K. But $h^{(p)}(x)$ is equal to zero on I_k, and is $F(x)(1 \pm a_i)$ on I_i, which is of constant sign. It follows that the functions $h^{(p-1)}, \cdots, h$ can change sign at most $2n - 1$ times on K, which is a contradiction. This establishes (7).

We see that $\Re$ contains the ball of radius $\| F_p \|$, with center in the origin, of the space X_{2n}. Relation (6) follows from Theorem 2. ∎

In particular, let $\omega(h)$, $0 \leqslant h \leqslant \pi$ be a concave modulus of continuity, and let $F(x)$ be the odd, $(2\pi/n)$-periodic function given by 8.5(6). We know from 3.5(9) that the functions F and $f = \sum a_i g_i$, $| a_i | \leqslant 1$ belong to Λ_ω. Therefore,

$$d_{2n-1}(\Lambda_{p\omega}) \geqslant \| F_p \| . \tag{8}$$

For $p = 0$, we have $\| F_0 \| = \frac{1}{2} \omega(\pi/n)$, and from the inequality 8.5(3), we obtain

THEOREM 5. The trigonometric functions 1, $\cos x$, $\sin x$, $\cdots$, $\cos (n - 1) x$, $\sin (n - 1) x$ are an extremal system for the class $\Lambda_\omega[K]$, where ω is a concave modulus of continuity, and

$$d_{2n-1}(\Lambda_\omega) = \frac{1}{2} \omega \left(\frac{\pi}{n} \right). \tag{9}$$

If $F(x)$ is the odd $(2\pi/n)$-periodic function, defined by $F(x) = 1$, $0 \leqslant x \leqslant \pi/n$, then $F_p \in W_p^*$, and also all functions $f = \sum a_i g_i$, $| a_i | \leqslant 1$, have pth integrals f_p belonging to W_p^*. From Theorems 4 and 5, Chapter 8, we obtain

THEOREM 6 (Tihomirov [93]). The functions $1, \cdots, \sin (n - 1) x, \cos (n - 1) x$ are an extremal system for W_p^*, and

$$d_{2n-1}(W_p^*) = K_p n^{-p}, \qquad p = 1, 2, \cdots, \tag{10}$$

where the constants K_p are given by 8.3(2).

The next two theorems give partial answers to some natural questions.

Let $A' = A'(\rho)$, $\rho > 1$ be the subset of $C[-1, +1]$ that consists of all functions on $[-1, +1]$, which have an analytic extension, bounded by 1, into the ellipse E_ρ.

THEOREM 7

$$\rho^{-n} \leqslant d_n(A') \leqslant 2(\rho - 1)^{-1} \rho^{-n}. \tag{11}$$

Proof. Let $X_{n+1} \subset C[-1, +1]$ consist of all polynomials P_n of degree n. If $\| P_n \| \leqslant \rho^{-n}$, then by Theorem 7, Chapter 3, $| P_n(z) | \leqslant 1$ in E_ρ. We see that the ball of radius ρ^{-n} and with center in the origin in X_{n+1} is contained in A'. This gives the left-hand inequality (11). The right-hand inequality follows from the estimate 5.5(6) of $E_n(f)$, $f \in A'$, obtained in the proof of Theorem 8, Chapter 5. ∎

A similar argument (which uses Theorem 12, Chapter 8 and the inequalities 3.2(6) and that of Problem 2, Chapter 3) shows that for the class $A_p(\rho)$, $p = 0, 1, \cdots, \rho > 1$ defined in Chapter 8, Sec. 7, one has

$$d_n(A_p(\rho)) = \frac{\rho^{p-n}}{n(n-1)\cdots(n-p+1)}, \qquad n \geq p. \tag{12}$$

THEOREM 8. If $p = 1, 3, \cdots$ is odd, then for $W_p[-\pi, \pi]$,

$$d_{2n-1}(W_p) = K_p n^{-p}(1 + O(n^{-1})). \tag{13}$$

Proof. We need a generalization of a part of Theorem 4, Chapter 8. We consider continuous functions $f(x)$ on K, with absolutely continuous $f^{(p-1)}(x)$, for which

$$|f^{(p)}(x) + b| \leq 1 \qquad \text{a.e.} \qquad \text{for some constant } b. \tag{14}$$

For all such functions f, $E^*_{n-1}(f) \leq K_p n^{-p}$.

Reexamining the proof of Theorems 4 and 5, Chapter 8, we see that if $p = 1, 3, \cdots$, then both the function $H = D_p$ [8.3(10)] and its polynomial of best L^1-approximation S_{n-1} (See Theorem 2, Chapter 8) are odd. Using the relations 8.2(6) and 8.3(8), and the function f_{nH} [8.3(6)], we obtain, similarly to 8.3(9),

$$E^*_{n-1}(f) \leq \max_x \left| \int_K [H(x-t) - S_{n-1}(x-t)]f^{(p)}(t)\,dt \right|$$

$$= \max_x \left| \int_K [H(t) - S_{n-1}(t)]\,[f^{(p)}(x-t) + b]\,dt \right|$$

$$\leq \int_K |H(t) - S_{n-1}(t)|\,dt \leq \left| \int_K H(t)\, \text{sign} \sin nt\, dt \right|$$

$$= |f_{nH}(0)| \leq E^*_{n-1}(W^*_p).$$

Now let g be a function of the class $W_p[-\pi, \pi]$, not necessarily periodic; we have $|g^{(p)}(x)| \leq 1$ a.e. If $b = (2\pi)^{-1}\int_{-\pi}^{\pi} g^{(p)}dx$, then the difference $g^{(p)} - b$ has a periodic pth integral f, and the function f satisfies (14). Hence, $E^*_{n-1}(f) \leq K_p n^{-p}$. On the other hand, $g - f$ is an algebraic polynomial of degree p. It follows that the function g is approximable, with an error not exceeding $K_p n^{-p}$, by a linear combination of the $p + 2n - 1$ functions $1, x, \cdots$, x^p, $\cos x$, $\sin x$, $\cdots$, $\cos(n-1)x$, $\sin(n-1)x$. Hence, $d_{p+2n-1}(W_p) \leq K_p n^{-p}$, and $d_{2n-1}(W_p) \leq K_p n^{-p}(1 + O(n^{-1}))$. But W_p may be regarded as a subset of W_p, so that $d_{2n-1}(W_p) \geq d_{2n-1}(W^*_p) = K_p n^{-p}$. ∎

5. Differential Operators

In this section we shall collect, for the most part without proofs, some basic facts about linear differential operators, which will be needed in the next section.

(The reader will find the proofs, for example, in [27]). Let $L^2 = L^2[a, b]$ be the space of square integrable real functions f on $[a, b]$, with norm

$$\|f\| = \left(\int_a^b |f(x)|^2 \, dx \right)^{1/2}.$$

Let $L^{2,n}$ denote the subspace of L^2, which consists of all functions f that have an absolutely continuous derivative $f^{(n-1)}$, with $f^{(n)} \in L^2$.

We shall consider *linear differential expression* of order $n \geqslant 1$:

$$l(y) = p_0(x) \, y^{(n)} + p_1(x) \, y^{(n-1)} + \cdots + p_n(x) \, y. \tag{1}$$

Here, $p_k(x)$, $k = 0, \cdots, n$ are continuous real functions, defined on $[a, b]$, and $p_0(x)$ does not vanish. A differential expression $l^*(z)$ is *adjoint* to $l(y)$ if

$$\int_a^b l(y) \, z \, dx = V(y, z) + \int_a^b y l^*(z) \, dx, \qquad y, z \in L^{2,n}, \tag{2}$$

where $V(y, z)$ is a bilinear form of the values of the derivatives of y and z of orders $k = 0, \cdots, n - 1$ at a and b. More exactly, $V(y, z)$ should be of the form $V(b) - V(a)$, where

$$V(x) = \sum_{k+l < n} \alpha_{kl} y^{(k)}(x) \, z^{(l)}(x)$$

and α_{kl} are constants. If the adjoint expression exists, it is unique. If the $p_k(x)$ have continuous derivatives of order $n - k$, $k = 0, \cdots, n$, l^* exists and can be found by integration by parts. From

$$\int_a^b p_{n-k} z y^{(k)} \, dx = [p_{n-k} z y^{(k-1)} - (p_{n-k} z)' y^{(k-2)}$$

$$+ \cdots + (-1)^{k-1} (p_{n-k} z)^{(k-1)} y]_{x=a}^{x=b}$$

$$+ (-1)^k \int_a^b y (p_{n-k} z)^{(k)} \, dx$$

for $k = 0, \cdots, n$, we obtain (2) by summation, with V of the required form and

$$l^*(z) = (-1)^n (p_0 z)^{(n)} + (-1)^{n-1} (p_1 z)^{(n-1)} + \cdots + p_n z. \tag{3}$$

A special case of the formula (2) is

$$\int_a^b l(y) \, z \, dx = \int_a^b y l^*(z) \, dx \qquad \text{if } y, z \in L^{2,n}$$

$$\text{and} \quad z(x) = z'(x) = \cdots = z^{(n-1)}(x) = 0 \qquad \text{for } x = a, b. \tag{4}$$

The differential expression $l(y)$ is *self-adjoint* if $l^*(y)$ is identical with $l(y)$.

We shall assume from now on that the functions p_k in (1) have n continuous derivatives. In this case, we can form the differential expression of order $2n$:

$$L(y) = l^*(l(y)). \tag{5}$$

We show that L is *self-adjoint*. We have, using (2) two times, if $y, z \in L^{2,2n}$,

$$\int_a^b L(y) z \, dx = V_1(y, z) + \int_a^b l(y) l(z) \, dx$$

$$= V_1 + V_2(y, z) + \int_a^b y L(z) \, dx. \tag{6}$$

Here, V_1 is a bilinear form of the derivatives of orders $0, \cdots, n-1$ of z and of $l(y)$ at a and b, and V_2 is a similar form, with z and y interchanged. Hence, $V_1 + V_2$ is a bilinear form of the derivatives of y and z of orders $0, \cdots, 2n-1$ at a and b.

If $y_0 \in L^2$ is given, $l(y) = y_0$ is a differential equation of order n, which has solutions $y \in L^{2,n}$ for each $y_0 \in L^2$. If $y_0 = 0$, then each of the solutions y belongs to $L^{2,2n}$. This is obtained by solving the equation $l(y) = 0$ for $y^{(n)}$ and by differentiating the resulting relation $n-1$ times.

We consider the expression (1) and certain *boundary value conditions*:

$$B_k(y) = \alpha_{k0} y^{(n-1)}(a) + \cdots + \alpha_{k,n-1} y(a) + \beta_{k0} y^{(n-1)}(b) + \cdots + \beta_{k,n-1} y(b) = 0,$$

$$k = 0, \cdots, n-1. \tag{7}$$

Together, $l(y)$ and the conditions (7) define a *differential operator* $A(y)$, which has values $l(y)$ and is defined for all $y \in L^{2,n}$ that satisfy (7).

The *domain of definition* $\mathfrak{D}(A)$ of A consists of all these functions y. The differential operator A is *self-adjoint* if the differential expression $l(y)$ is self-adjoint and if

$$\int_a^b A(y) z \, dx = \int_a^b y A(z) \, dx, \qquad y, z \in \mathfrak{D}(A). \tag{8}$$

LEMMA 4. The differential operator A that corresponds to a differential expression (5) and the boundary value conditions

$$l(y) = [l(y)]' = \cdots = [l(y)]^{(n-1)} = 0, \qquad \text{for } x = a, b \tag{9}$$

is self-adjoint.

This follows from the proof of the formula (6). If the functions y and z both satisfy (9), then $V_1 = V_2 = 0$. ∎

A function $f \in \mathfrak{D}(A)$ is an *eigenfunction* of the operator A if $f \neq 0$ and if $Af = \lambda f$ for some complex number λ; this λ is called an *eigenvalue* of A (corresponding to f). The number of linearly independent functions f belonging to an eigenvalue λ is the *multiplicity* of λ. We shall need the deep and important theorems (see, for example, [27]) about the eigenvalues of a self-adjoint differential operator A. The operator A has countably many eigenvalues, λ_n, and $|\lambda_n| \to +\infty$; all eigenvalues are real and have finite multiplicity. The eigenfunctions of A can be assumed to form a complete orthonormal system in $L^2[a, b]$.

The eigenvalues can rarely be found explicitly. But often *asymptotic formulas* for the λ_k exist. Very good theorems of this kind are due to Birkhoff [36]. (They are reproduced in [27].) We need the following special case:

Let $L(y)$ be a differential expression of degree $2n$, and let the boundary conditions be given by

$$\left.\begin{array}{l} \alpha_{k0}y^{(n+k)}(a) + \alpha_{k1}y^{(n+k-1)}(a) + \cdots + \alpha_{k,n+k}y(a) = 0 \\ \beta_{k0}y^{(n+k)}(b) + \beta_{k1}y^{(n+k-1)}(b) + \cdots + \beta_{k,n+k}y(b) = 0 \end{array}\right\} \; k = 0, \cdots, n-1, \quad (10)$$

with $\alpha_{k0} \neq 0$, $\beta_{k0} \neq 0$. [In particular, conditions (9) have this form.] Then the operator A has infinitely many eigenvalues, all but finitely many of which are simple; and for $k \to \infty$,

$$\lambda_k = \left(\frac{\pi}{b-a}\right)^{2n} k^{2n}[1 + O(k^{-1})]. \cdot \quad (11)$$

6. Widths of the Set $\mathfrak{K}_l$

We can now discuss a problem, solved by Kolmogorov ([60]; he treats the case $l(y) = y^{(n)}$). Let $l(y)$ be a differential expression of the type 4(1). If $\mathfrak{K}_l$ is the set of $f \in L^{2,n}$, for which $\|\, l(f)\,\|_2 \leqslant 1$, what are the widths of $\mathfrak{K}_l$ in L^2? The (generalized) Kolmogorov's theorem is

THEOREM 9. If λ_k are the eigenvalues of the differential operator A given by the expression $L(y) = l^*(l(y))$ and the boundary value conditions 4(9), then $\lambda_k \geqslant 0$. If the λ_k are arranged in order of increasing magnitude and repeated each according to its multiplicity, then

$$d_k(\mathfrak{K}_l) = \frac{1}{\sqrt{\lambda_{k+1}}}, \qquad k = 0, 1, \cdots. \quad (1)$$

In particular, $d_k(\mathfrak{K}_l) = +\infty$ for $k = 0, \cdots, n-1$ and

$$d_k(\mathfrak{K}_l) = \left(\frac{b-a}{\pi}\right)^n k^{-n}[1 + O(k^{-1})], \qquad k \geqslant n.$$

Proof. If y is a solution of $l(y) = 0$, then $y \in L^{2,2n}$ and y satisfies 4(9) and $A(y) = 0$. Conversely, if $L(y) = 0$ and if y satisfies 4(9) for $x = a$, then $z = l(y) = 0$ follows from the uniqueness theorem for the differential equation

$l^*(z) = 0$. This establishes that the eigenfunctions of A corresponding to $\lambda = 0$ are exactly the solutions of $l(y) = 0$, and consequently that $\lambda = 0$ has multiplicity n.

Let λ_k $(\lambda_1 = \cdots = \lambda_n = 0)$, $k = 1, 2, \cdots$, be all eigenvalues of A, and let $\psi_k \in L^{2,n}$ be the corresponding eigenfunctions, which form a complete orthonormal system in L^2. Each $f \in L^2$ has a representation $f = \sum_1^\infty a_k \psi_k$, $\sum a_k^2 < +\infty$.

We put $\phi_k = l(\psi_k)$. Then $\phi_1 = \cdots = \phi_n = 0$. From 4(4) we derive

$$\int_a^b \phi_k \phi_l dx = \int_a^b l(\psi_k)\, l(\psi_l)\, dx = \int_a^b \psi_k L(\psi_l)\, dx$$

$$= \lambda_l \int_a^b \psi_k \psi_l\, dx$$

$$= \begin{cases} 0, & \text{if} \quad k = l; \\ \lambda_k, & \text{if} \quad k \neq l. \end{cases} \tag{2}$$

This shows that $\lambda_k \geqslant 0$, $k = 1, 2, \cdots$. It follows also that $\phi_k/\sqrt{\lambda_k}$, $k = n+1, \cdots$, is an orthonormal system on $[a, b]$.

For $f \in L^{2,n}$ we have, by 4(4),

$$\int_a^b l(f) \phi_k\, dx = \int_a^b f L(\psi_k)\, dx = \lambda_k \int_a^b f \psi_k\, dx, \qquad k = 1, 2, \cdots. \tag{3}$$

Let $g \in L^2$ be given, and let f be a solution of $l(f) = g$. If we assume that $\int_a^b g \phi_k\, dx = 0$, $k > n$, We obtain from (3) that $\int_a^b f \psi_k\, dx = 0$, $k > n$, so that f is a linear combination of $\psi_1, \cdots, \psi_n$. Therefore, $g = l(f) = 0$. We see that the system $\phi_k/\sqrt{\lambda_k}$ is complete.

We can now prove that $f = \sum a_k \psi_k \in L^2$ belongs to $\mathfrak{R}_l$ if and only if

$$\sum_{k=1}^\infty \lambda_k a_k^2 \leqslant 1. \tag{4}$$

First, $f \in \mathfrak{R}_l$ implies that $l(f) = \sum_{n+1}^\infty b_k \phi_k/\sqrt{\lambda_k}$ with $\sum b_k^2 \leqslant 1$, and from (3), $b_k = \sqrt{\lambda_k}\, a_k$. Hence, (4) must hold. On the other hand, if (4) holds, then the function $g = \sum_{n+1}^\infty \sqrt{\lambda_k}\, a_k(\phi_k/\sqrt{\lambda_k})$ belongs to L^2 and $\| g \| \leqslant 1$. Let $f_0 = \sum a_k' \psi_k \in L^{2,n}$ be some solution of $l(f_0) = g$. By (3), $a_k = a_k'$, $k > n$. Thus, $l(f - f_0) = 0$; hence, $l(f) = g$, $f \in \mathfrak{R}_l$.

The space $L^2[a, b]$ is isometric with the space l^2 of Fourier coefficients a_k, and the set $\mathfrak{R}_l$ in L^2 is isometric to the ellipsoid D given by 4(3) with $\delta_k = 1/\sqrt{\lambda_k}$. Therefore, (1) is a consequence of Theorem 4.   ∎

7. Notes

1. Theorem 8 is due to Tihomirov [93], who states it for $p = 1, 2, \cdots$. However, his proof is adequate only for odd p. In Tihomirov [95] an exact formula for $d_n(W_p)$ for all $p = 1, 2, \cdots$ is announced.

2. Theorem 8 is of considerable interest from the following point of view. An asymptotic expression for the degree of approximation of the class $W_p[-\pi, \pi]$, $p = 1, 2, 3, \cdots$, by algebraic polynomials has been obtained by Bernstein [2, vol. II, pp. 413-415]: $E_n(W_p) \sim \pi^p K_p n^{-p}$. (This corresponds to the exact formula of Favard, $E_n(W_p^*) = K_p n^{-p}$). On the other hand, 4(13), Theorem 8, gives us $d_n(W_p) \sim 2^p K_p n^{-p}$. This shows that for $p = 1, 3, \cdots$, the algebraic polynomials are not the best means of approximation of the class W_p: There exist systems that give about $(\pi/2)^p$ times better approximation.

3. Theorem 9 applies also to complex-valued differential expressions $l(y)$. In this case, the functions ψ_n, ϕ_n are complex, and $d_n(\Re_l)$ denotes the nth width of $\Re_l$ in the complex space L^2.

Using an interesting new approach, Mitjagin [81] computes widths of classes of periodic functions of several variables.

PROBLEMS

1. If $\bar{A}$ is the closure of A and a is real, then $d_n(\bar{A}) = d_n(A)$, $d_n(aA) = |a| d_n(A)$.

2. If A is a subset of X, which is symmetric with respect to the origin, that is, if $x \in A$ implies $-x \in A$, then

$$d_n(A) = \inf_{Y_n} E_{Y_n}(A).$$

Here, Y_n are all n-dimensional manifolds of X, that is, all translations of all n-dimensional subspaces X_n of X.

3. By approximating functions $f \in \Lambda_\omega^*$ by step functions, show that $d_n^M(\Lambda_\omega^*) \leqslant \omega(\pi/n)$. ($M$ is the Banach space of all bounded measurable functions on K.)

4. Lemma 1 holds if the space $C[B]$ is replaced by the space M of all bounded functions on B.

5. Using Problem 4 and the method of Problem 3, show that $d_n(\Lambda_\omega^1[B]) = \omega(1/n)$, where $B = [0, 1]$.

6. Show that if X is a subspace of the Banach space X', then $d_n^{X'}(A) \leqslant d_n^X(A)$.

7. In Problem 6, it can happen that $d_n^{X'}(A) < d_n^X(A)$. An example is provided by Problem 3 and Theorem 5, but there are examples with a finite dimensional space X'. (Tihomirov).

8. Using the method of the proof of Theorem 8, show that

$$d_{2n-1}(W_p) \leqslant 2K_p n^{-p}, \qquad p = 1, 2, \cdots.$$

9. Theorem 9 (except for the simplicity of the eigenvalues) remains true in the periodic case. Here $\Re_l^*$ is the subset of $L^{2,n}[K]$, which consists of functions f on K with $\| l(f) \|_2 \leqslant 1$, and the λ_k are the eigenvalues of the differential equation $l^*(l(y)) = \lambda y$ (*without* boundary value conditions).

10. If $l(y) = y^{(n)}$ in Problem 9, then

$$d_0(\Re_l^*) = +\infty, \qquad d_{2k-1}(\Re_l^*) = d_{2k}(\Re_l^*) = k^{-n}, \qquad k = 1, 2, \cdots.$$

Prove this directly.

[10]

Entropy

1. Entropy and Capacity

In Chapter 9 we have discussed, for a given subset A of a Banach space X, the sequence $d_n(A)$ of widths of A. For a compact set A, $d_n(A) \to 0$ for $n \to \infty$ (Chapter 9, Sec. 1); the rapidity of convergence of $d_n(A)$ to zero describes to some extent the "size" of A. The widths $d_n(A)$ are small if A deviates little from certain finite-dimensional subsets of X. We shall now study another invariant, the entropy of a set A. It also characterizes the "size" or the "massiveness" of A, but depends less than $d_n(A)$ upon the shape of A.

We begin by recalling some well-known definitions. Let A be a subset of a metric space X, and let $\epsilon > 0$ be given.

(a) A family $U_1, \cdots, U_n$ of subsets of X is called an ϵ-*covering* of A if the diameter of each U_k does not exceed 2ϵ and if the sets U_k cover A: $A \subset \bigcup_1^n U_k$.

(b) A finite set of points $x_1, \cdots, x_p$ of X is called an ϵ-*net* for A if for each $x \in A$ there is at least one point x_k of the net at a distance from x not exceeding ϵ : $\rho(x, x_k) \leqslant \epsilon$.

(c) Points $y_1, \cdots, y_m$ of A are called ϵ-*distinguishable* if the distance between each two of them exceeds ϵ: $\rho(y_i, y_k) > \epsilon$ for all $i \neq k$.

There are relations between these definitions. If $x_1, \cdots, x_p$ is an ϵ-net for A, then according to (a), (b), there is also an ϵ-covering of A that consists of p sets; for the U_k we can take the closed balls with centers x_k and radii ϵ. A standard theorem of topology [28, p. 123] guarantees that each compact set A contains a finite ϵ-net for each $\epsilon > 0$. Hence, there is also a finite ϵ-covering for each $\epsilon > 0$. Moreover, a compact set A can contain only finitely many ϵ-distinguishable points. From now on, we shall suppose that A is compact.

Of course, for a given $\epsilon > 0$, the number n of sets U_k in a covering family (a) depends on the family, but the minimal value of n, $N_\epsilon(A) = \min n$ is an invariant of the set A, which depends only upon $\epsilon > 0$. The logarithm

$$H_\epsilon(A) = \log N_\epsilon(A) \tag{1}$$

is the *entropy* (or *metric entropy*, in contradistinction to the probabilistic entropy) of the set A. Kolmogorov ([62]; see also Kolmogorov and Tihomirov [65]) conceived the idea of characterizing the "massiveness" of a set A by means

150

of the function (1). We are interested, of course, in the asymptotic behavior of $H_\epsilon(A)$ for $\epsilon \to 0$; in general, $H_\epsilon(A)$ will increase rapidly to infinity. We take logarithms in (1) partly because $N_\epsilon(A)$ is often so large that it cannot be dealt with conveniently.

In the same way for a given $\epsilon > 0$, numbers p, m in (b) and (c) depend on the choice of points, but $P_\epsilon(A) = \min p$ and $M_\epsilon(A) = \max m$ are invariants of the set A (we shall see in a moment that $M_\epsilon(A)$ exists). The expressions

$$H_\epsilon^X(A) = \log P_\epsilon(A), \qquad C_\epsilon(A) = \log M_\epsilon(A) \tag{2}$$

are called, respectively, the *entropy* of A *with respect* to X, and the *capacity* of A. It is easy to see that $H_\epsilon(A)$, $C_\epsilon(A)$ depend only on the compact metric space A itself (and not on the larger space X in which A is contained), but $H_\epsilon^X(A)$ may depend on X. All these functions are monotone-increasing in A and decreasing in ϵ.

The main general result about entropies is the simple

THEOREM 1. *For each $\epsilon > 0$, and each compact set A,*

$$C_{2\epsilon}(A) \leqslant H_\epsilon(A) \leqslant H_\epsilon^X(A) \leqslant C_\epsilon(A). \tag{3}$$

Proof. Each p of (b) is also an n of (a); hence, $N_\epsilon(A) \leqslant P_\epsilon(A)$. Further, if $y_1, \cdots, y_m$ are some 2ϵ-distinguishable points of A, and $U_1, \cdots, U_n$ is an ϵ-covering of A, then $m \leqslant n$, for otherwise two different points y_i will be contained in the same set U_k. Thus, $M_{2\epsilon}(A) \leqslant N_\epsilon(A)$; we see also that $M_\delta(A)$ is defined for each $\delta > 0$. Finally, if $y_1, \cdots, y_m$ are ϵ-distinguishable points of A, $m = M_\epsilon(A)$ in number, then they also form an ϵ-net for A; hence, $P_\epsilon(A) \leqslant M_\epsilon(A)$. Taking logarithms, we obtain (3). ∎

For example, if A is the interval $[a, b]$ with the usual metric, then $N_\epsilon(A)$ is the smallest n with $n \geqslant l/(2\epsilon)$, $l = b - a$, and $M_\epsilon(A)$ is the largest m with $m < l/\epsilon$; hence

$$N_\epsilon(A) = \frac{l}{2\epsilon} + O(1), \qquad M_\epsilon(A) = \frac{l}{\epsilon} + O(1). \tag{4}$$

This gives us $H_\epsilon(A)$, $C_\epsilon(A)$. In the most interesting cases we cannot determine the entropy of A with similar precision; we may be content with finding it only up to a strong or a weak equivalence: $H_\epsilon(A) \sim \phi(\epsilon)$ or $H_\epsilon(A) \approx \phi(\epsilon)$, where $\phi(\epsilon)$ is a known function. For this purpose, (3) is used in the following way: one seeks an upper bound for $H_\epsilon(A)$ (or $H_\epsilon^X(A)$) and a lower bound for $C_{2\epsilon}(A)$; if the bounds are close to each other, a good estimate for both entropy and capacity results.

A set $U \subset X$ of diameter 2ϵ is called *centerable* (in X) if U has a center x_0, that is, a point $x_0 \in X$ such that $\rho(x, x_0) \leqslant \epsilon$ for all $x \in U$. If all sets U_k in (a) are centerable, then their centers form an ϵ-net for A. In what follows, we shall be mainly concerned with the determination of entropies of subsets of spaces of continuous functions $X = C[B]$, where B is a compact metric space. In this connection, the following remark is of interest: each compact subset A of $C[B]$ is centerable.

Indeed, by Arzelà's theorem, there is a function $\omega(\delta) > 0$, which tends to zero with δ, so that

$$| f(x) - f(x') | \leqslant \omega(\delta) \qquad \text{for} \qquad \rho(x, x') \leqslant \delta, \quad f \in A. \tag{5}$$

Also the functions $\overline{f}(x) = \sup_{f \in A} f(x)$, $\underline{f}(x) = \inf_{f \in A} f(x)$ satisfy (5), and hence belong to $C[B]$. The function $f_0(x) = \frac{1}{2}\{\underline{f}(x) + \overline{f}(x)\}$ is a center of A.

In the definition (a) we may assume that the sets U_k are closed and subsets of A; hence that they are compact. It follows, therefore, if A is a compact subset of $X = C[B]$:

$$H_\epsilon^X(A) = H_\epsilon(A) \qquad \text{if} \qquad A \subset X = C[B]. \tag{6}$$

The entropy of a cartesian product can be estimated if the entropies of all factors are known. Assume for example that $A = \prod_1^s A_k$ is a subset of the s-dimensional euclidean space R_s, and that each A_k is a compact subset of the kth coordinate axis. Let $\epsilon > 0$ be given, and let $\epsilon_1 = \epsilon/\sqrt{s}$. For each k, let the sets U_{ik}, $i = 1, \cdots, N_{\epsilon_1}(A_k)$ be a minimal ϵ_1-covering of A_k. Then all possible products $\prod_{k=1}^s U_{i_k k}$ with $1 \leqslant i_k \leqslant N_{\epsilon_1}(A_k)$ are $\prod_1^s N_{\epsilon_1}(A_k)$ in number, have diameters $\leqslant 2\epsilon$ and form an ϵ-covering of A. Hence, $N_\epsilon(A)$ does not exceed the last product, and

$$H_\epsilon(A) \leqslant \sum_{k=1}^s H_{\epsilon/\sqrt{s}}(A_k). \tag{7}$$

In the same way we obtain

$$C_\epsilon(A) \geqslant \sum_{k=1}^s C_\epsilon(A_k). \tag{8}$$

In particular, if each A_k is an interval, then by (4), both $C_\epsilon(A_k)$ and $H_\epsilon(A_k)$ are $\log(1/\epsilon) + O(1)$, and we obtain for an s-dimensional parallelepiped A in R_s:

$$C_\epsilon(A) \text{ and } H_\epsilon(A) = s \log \frac{1}{\epsilon} + O(1), \qquad \epsilon \to 0. \tag{9}$$

Because of the monotony of our set functions, we have, more generally:

THEOREM 2.   Relation (9) holds for each bounded subset A of R_s with interior points.

For sets $A \subset R_s$ that satisfy some modest additional assumptions of regularity, it is not hard to prove (see [65]) somewhat more than (9), namely,

$$N_\epsilon(A) \sim \lambda_s \mid A \mid \epsilon^{-s},$$

$$M_\epsilon(A) \sim \mu_s \mid A \mid \epsilon^{-s}. \tag{10}$$

Here $\mid A \mid$ is the s-dimensional volume of A, and λ_s, μ_s are some positive constants (which depend on s, but not on A). However, the precise values of λ_s and μ_s, $s \geqslant 3$, are not known. The determination of λ_s, μ_s is equivalent to the problems of finding the most economical covering of the space R_s by balls of radius 1, and finding the tightest packing of balls into this space.

2. Sets of Continuous and Differentiable Functions

Let B be an s-dimensional parallelepiped in the euclidean space R_s, ω a modulus of continuity, and p an integer, $p = 0, 1, \cdots$. Then

$$\Lambda = \Lambda^s_{p\omega}(M_0, \cdots, M_{p+1}; B)$$

is the set of all continuous functions $f(x) = f(\xi_1, \cdots, \xi_s)$ on B that have continuous partial derivatives of orders $\leqslant p$. The partial derivatives of order k must satisfy $\mid D^k f(x) \mid \leqslant M_k$, $k = 0, \cdots, p$, while those of order p have, in addition, moduli of continuity not exceeding $M_{p+1}\omega(h)$. (See Chapter 3, Sec. 7.)

Let $\delta = \delta(\epsilon)$ be defined by the equation

$$\delta^p \omega(\delta) = \epsilon. \tag{1}$$

[If, for some $\epsilon > 0$, (1) has several roots δ, we can interpret $\delta(\epsilon)$ to be any of them.]

THEOREM 3. For each Λ, there are constants K, L, β, γ for which (for all sufficiently small $\epsilon > 0$)

$$\frac{K}{\delta(\beta\epsilon)^s} \leqslant C_{2\epsilon}(\Lambda) \leqslant H_\epsilon(\Lambda) \leqslant \frac{L}{\delta(\gamma\epsilon)^s}. \tag{2}$$

In particular, if $\omega(h) = h^\alpha$, $0 < \alpha \leqslant 1$, then $\delta(\epsilon) = \text{const } \epsilon^{1/(p+\alpha)}$ and

$$C_\epsilon(\Lambda) \approx H_\epsilon(\Lambda) \approx \left(\frac{1}{\epsilon}\right)^{s/(p+\alpha)} \tag{3}$$

Proof. (a) We estimate $H_\epsilon(\Lambda)$ from above. We consider in detail only the case $s = 1$; B is then a closed interval $[a, b]$ of length $l = b - a$. The main tool at our disposal for the functions $f \in \varLambda$ is the Taylor formula

$$f^{(k)}(x') = f^{(k)}(x) + \frac{x' - x}{1!} f^{(k+1)}(x) + \cdots + \frac{(x' - x)^{p-k}}{(p - k)!} f^{(p)}(x) + R_k(f),$$

$$R_k(f) = \frac{(x' - x)^{p-k}}{(p - k)!} [f^{(p)}(\xi) - f^{(p)}(x)], \qquad k = 0, \cdots, p, \tag{4}$$

where ξ is between x and x'. Let $x' - x$ be small. Suppose that we want to calculate with a prescribed accuracy the values $f(x'), \cdots, f^{(p)}(x')$ by means of (4). What margin of error could we allow for each of the values $f(x), \cdots, f^{(p)}(x)$? It is clear that we should allot higher margins of error to higher derivatives: This will be counterbalanced by high powers of $x' - x$ contained in the corresponding terms of (4).

For a given $\epsilon > 0$, we associate with $f^{(k)}$ the number

$$\epsilon_k = \delta^{p-k}\omega(\delta), \qquad k = 0, \cdots, p. \tag{5}$$

LEMMA 1. Let f_1, f_2 be two functions, defined on $[a, b]$, that belong to the class $\varLambda$. If their derivatives satisfy at the point x the inequalities

$$|f_1^{(l)}(x) - f_2^{(l)}(x)| \leqslant \epsilon_l, \qquad l = k, \cdots, p,$$

then

$$|f_1^{(k)}(x') - f_2^{(k)}(x')| \leqslant 4\epsilon_k \qquad \text{for } |x' - x| \leqslant \delta, \ k = 0, \cdots, p. \tag{6}$$

To prove this, we apply (4) to the function $g = f_1 - f_2$:

$$|R_k(g)| \leqslant \frac{2\delta^q}{q!} \omega(\delta) = \frac{2}{q!} \epsilon_k, \qquad q = p - k,$$

$$\left| \frac{(x' - x)^m}{m!} g^{(k+m)}(x) \right| \leqslant \frac{\delta^m}{m!} \epsilon_{k+m} = \frac{1}{m!} \epsilon_k, \qquad m = 0, \cdots, q,$$

so that

$$|g^{(k)}(x')| \leqslant \left\{ \sum_{m=0}^{q} \frac{1}{m!} + \frac{2}{q!} \right\} \epsilon_k \leqslant 4\epsilon_k. \qquad \blacksquare$$

For a given $\epsilon > 0$, and $\delta = \delta(\epsilon)$ defined by (1), we shall construct an economical 2ϵ-covering of $\varLambda$ by sets U. The $N + 1$ points $x_0 = a$, $x_1 = a + \delta$, $\cdots$, $x_N = a + N\delta$, $N = [l/\delta]$ divide $[a, b]$ into intervals of lengths $\leqslant \delta$. Each set U will consist of functions $f \in \varLambda$ for which the values of the derivatives $f^{(k)}$ at the points x_i are fixed within errors not exceeding ϵ_k. More formally, each U will be given by a matrix (m_{ki}), $k = 0, \cdots, p$, $i = 0, \cdots, N$, with integral entries m_{ki}. A function $f \in \varLambda$ belongs to U if and only if

$$m_{ki}\epsilon_k \leqslant f^{(k)}(x_i) < (m_{ki} + 1) \epsilon_k, \qquad k = 0, \cdots, p; \qquad i = 0, \cdots, N. \tag{7}$$

Each point $x \in B$ is at a distance $< \delta$ from one of the points x_i. It follows from Lemma 1 for $k = 0$ that if two functions f_1, f_2 belong to the same U, then $\|f_1 - f_2\| \leqslant 4\epsilon$, so that the sets U form a 2ϵ-covering of Λ.

It remains to estimate the number of U that are not empty, that is, contain points of Λ. We first count the number of possible first columns m_{k0}, $k = 0, \cdots, p$ of our matrices. Let k be fixed. Since all values $f^{(k)}(x_0)$ are in the intervals $[- M_k, M_k]$, there may exist at most $2(M_k/\epsilon_k + 1) \leqslant \text{const } \epsilon^{-1}$ nonempty intervals $[m_{k0}\epsilon_k, (m_{k0} + 1)\epsilon_k]$ and hence at most that many entries m_{k0}. Altogether, there are at most const ϵ^{-p-1} possible first columns $\{m_{k0}\}$. On the other hand, Lemma 1 tells us that if the m_{k0}, $k = 0, \cdots, p$ are given, there can be at most five possible values m_{k1} for each k. The number of possible second columns $\{m_{k1}\}$ for a given first column, is, therefore, at most 5^{p+1}. Continuing in this way, we see that the total number of nonempty sets U is at most const $\epsilon^{-p-1}(5^{p+1})^N$, and therefore if $\epsilon > 0$ is small,

$$H_{2\epsilon}(\Lambda) \leqslant \text{const } \left(\log \frac{1}{\epsilon} + N\right) = \text{const } \left(\log \frac{1}{\epsilon} + \delta(\epsilon)^{-1}\right) \leqslant L\delta(\epsilon)^{-1}.$$

This proves that

$$H_\epsilon(\Lambda) \leqslant L\delta(\tfrac{1}{2}\epsilon)^{-1}.$$

In case $s > 1$, the proof is similar, and we can omit the details. We can again use the Taylor expansion of the functions $f(x) = f(\xi_1, \cdots, \xi_s)$ from Λ and have Lemma 1 for their partial derivatives, with some constant γ instead of 4 in (5), and with the same ϵ_k. This time we can find $N \approx \delta^{-s}$ points $x_0, \cdots, x_N$ in B that form a δ-net and have the property that each point x_i, $i > 0$, is at a distance $\leqslant \delta$ from some point x_j, $j < i$. The covering sets U are constructed as before by prescribing with errors ϵ_k the values of all kth partial derivatives of f at the points x_i.

(b) To estimate $C_\epsilon(\Lambda)$ from below, we again define $\delta = \delta(\epsilon)$ by (1). The functions $f_0(x)$, given by 8.2(16) for this δ, belong to the class Λ. For two different f_0, f_0' among them, there is at least one value of i for which $f_0(z_i)$ and $f_0'(z_i)$ are of opposite sign. Therefore, $\|f_0 - f_0'\| \geqslant 2H(0) \geqslant 2m\epsilon$, $m > 0$. Thus we have found 2^n elements of Λ, which are $m\epsilon$-distinguishable. Since $n \sim \delta^{-s}$, we obtain

$$C_{m\epsilon}(\Lambda) \geqslant n \log 2 \geqslant \text{const } \delta(\epsilon)^{-s};$$

hence,

$$C_\epsilon(\Lambda) \geqslant \text{const } \delta(m^{-1}\epsilon)^{-s}. \qquad |$$

3. Entropy of Classes of Analytic Functions

We shall consider single-valued analytic functions of one or several complex variables. (For definition and properties see [25]). Let G be a bounded region in the s-dimensional complex space, $z = (z_1, \cdots, z_s) \in G$, a point of G

with complex coordinates z_j. We denote by $A^s(G)$ the set of all functions $f(\mathbf{z})$ defined and analytic on G. Clearly, $A^s(G)$ is a vector space over the complex numbers. If F is a compact subset of G, we denote by $A = A^s(F, G; M)$ the metric space of all functions $f \in A^s(G)$ that satisfy $|f(\mathbf{z})| \leqslant M$ for $\mathbf{z} \in G$; A is metrized by the supremum norm on F. It follows from the basic facts about analytic functions that all $f \in A^s(F, G; M)$ have on F uniformly bounded derivatives. Therefore, by Arzelà's theorem, the sets $A_s(F, G; M)$ are compact.

We shall adhere to the following notation in this section. Let $\mathbf{r} = (r_1, \cdots, r_s)$, $\mathbf{k} = (k_1, \cdots, k_s)$ denote points of R_s for which $r_j > 0$, $k_j = 0, 1, \cdots, j = 1, \cdots, s$. We write $\mathbf{x} > 0$ if $x_j > 0$, $\mathbf{x} < \mathbf{r}$ if $x_j < r_j$, $j = 1, \cdots, s$, $|\mathbf{z}| = (|z_1|, \cdots, |z_s|)$ $\mathbf{z}^{\mathbf{k}} = \prod_1^s z_j^{k_j}$, $\mathbf{z}^k = \prod_1^s z_j^k$ (if k is an integer), and $(\mathbf{x}, \mathbf{y}) = \sum_1^s x_j y_j$ for the scalar product. Thus, $|\mathbf{z} - \mathbf{a}| < \mathbf{r}$ represents the *polycylinder* D: $|z_j - a_j| < r_j$, $j = 1, \cdots, s$.

Functions $f(\mathbf{z})$, analytic in a polycylinder D: $|\mathbf{z}| < \mathbf{r}$ have an *absolutely convergent Taylor expansion*:

$$f(\mathbf{z}) = \sum_{\mathbf{k}} a_{\mathbf{k}} \mathbf{z}^{\mathbf{k}} = \sum_{k_1, \ldots, k_s = 0}^{\infty} a_{k_1 \ldots k_s} z_1^{k_1} \cdots z_s^{k_s}, \qquad \mathbf{z} \in D. \tag{1}$$

About the Taylor series (1) we note the following facts:

$$\sup_{|\mathbf{z}| < \mathbf{r}} |f(\mathbf{z})| \leqslant \sum_{\mathbf{k}} |a_{\mathbf{k}}| \mathbf{r}^{\mathbf{k}}, \tag{2}$$

and if f satisfies $|f(\mathbf{z})| \leqslant M$, $\mathbf{z} \in D$, then

$$|a_{\mathbf{k}}| \leqslant \frac{M}{\mathbf{r}^{\mathbf{k}}}; \tag{3}$$

the last inequality holds for all s-tuples

$$\mathbf{k} = (k_1, \cdots, k_s), \qquad k_j = 0, 1, \cdots, j = 1, \cdots, s.$$

A family $A^s(F, G; M)$, where F and G are two concentric polycylinders, F: $|\mathbf{z}| \leqslant \mathbf{r}$, G: $|\mathbf{z}| < \mathbf{r}'$, $r_j < r_j'$, $j = 1, \cdots, s$ will also be denoted by $A^s(\mathbf{r}, \mathbf{r}'; M)$. Essential here are the quotients $q_j = r_j'/r_j$ and the numbers $h_j = \log q_j$, $j = 1, \cdots, s$. Kolmogorov [62] proved that each set $A^s(F, G; M)$ has an entropy $\approx \log^{s+1}(1/\epsilon)$; see Theorem 6. In this section we discuss the special sets $A^s(\mathbf{r}, \mathbf{r}'; M)$, for which Vituškin [17] obtained more precise results.

THEOREM 4. For the family $A = A^1(r, r'; M)$ of analytic functions of one variable, with $q = r'/r$,

$$C_\epsilon(A) \text{ and } H_\epsilon(A) = \frac{1}{\log q} \log^2 \frac{1}{\epsilon} + O\left(\log \frac{1}{\epsilon} \log \log \frac{1}{\epsilon}\right). \tag{4}$$

More generally we have

THEOREM 5. For the set $A = A^s(\mathbf{r}, \mathbf{r}'; M)$,

$$C_\epsilon(A) \text{ and } H_\epsilon(A) = \frac{2}{(s+1)! \prod_1^s \log q_j} \log^{s+1} \frac{1}{\epsilon} + O\left(\log^s \frac{1}{\epsilon} \log \log \frac{1}{\epsilon}\right). \quad (5)$$

Proof of Theorem 5. (a) We estimate $H_\epsilon(A)$ from above, and for this purpose find an economical ϵ-covering of A by sets U. The plan of the construction is the following: For functions $f \in A$, we replace in (1) the terms with large $\mathbf{k}$ by zero; the remaining coefficients $a_\mathbf{k}$ we approximate by integral multiples of certain small numbers. Sets U will consists of functions f with identical approximations for the $a_\mathbf{k}$.

We shall need a simple lemma about multiple integrals. Let $\Delta(a, b)$, $0 \leqslant a < b \leqslant +\infty$ denote the region of the s-dimensional (real) euclidean space defined by $x_j \geqslant 0$, $j = 1, \cdots, s$, $a \leqslant x_1 + \cdots + x_s \leqslant b$. We denote the tetrahedron $\Delta(0, b)$ also by $\Delta(b)$.

LEMMA 2

$$\int \cdots \int_{\Delta(a)} dx_1 \cdots dx_s = \frac{a^s}{s!}; \quad (6)$$

$$\int \cdots \int_{\Delta(a,b)} f(x_1 + \cdots + x_s) \, dx_1 \cdots dx_s = \int_a^b f(y) \frac{y^{s-1}}{(s-1)!} \, dy. \quad (7)$$

Formula (6) gives the volume of $\Delta(a)$ and is obtained by induction in s; formula (7) follows from (6) after the substitution $y = x_1 \cdots + x_s, y_2 = x_2, \cdots$, $y_s = x_s$, for the point $(x_1, \cdots, x_s)$ is in $\Delta(a, b)$ if and only if $a \leqslant y \leqslant b$ and the point $(y_2, \cdots, y_s)$ is in the $(s-1)$-dimensional tetrahedron $\Delta(y)$. Let

$$N = \log \frac{1}{\epsilon} + (s-1) \log \log \frac{1}{\epsilon} + L, \quad (8)$$

where L is a positive constant, to be determined later. We define the region $T = T(N)$ of values of $\mathbf{k}$ by the inequality $(\mathbf{h}, \mathbf{k}) = h_1 k_1 + \cdots + h_s k_s \leqslant N$, where $\mathbf{h} = (h_1, \cdots, h_s)$, $h_j = \log q_j$, $j = 1, \cdots, s$. Let

$$s_N(f) = s_N(f, \mathbf{z}) = \sum_{\mathbf{k} \in T} a_\mathbf{k} \mathbf{z}^\mathbf{k}, \qquad f \in A. \quad (9)$$

We first show that, for a proper choice of L,

$$\|f - s_N(f)\| \leqslant \tfrac{1}{2} \epsilon, \qquad f \in A. \quad (10)$$

In fact, for $|\mathbf{z}| \leqslant \mathbf{r}$,

$$|f(\mathbf{z}) - s_N(f, \mathbf{z})| \leqslant \sum_{\mathbf{k} \notin T} |a_\mathbf{k}| \mathbf{r}^\mathbf{k}.$$

Hence, by (3) with $\mathbf{r}'$ instead of $\mathbf{r}$ and with $\mathbf{q} = (q_1, \cdots, q_n)$,

$$\| f - s_N(f) \| \leqslant M \sum_{\mathbf{k} \notin T}' \mathbf{q}^{-\mathbf{k}} = M \sum_{\mathbf{k} \notin T}' e^{-(\mathbf{h}, \mathbf{k})}. \tag{11}$$

Sums of this type are estimated by means of integrals (7). With each $\mathbf{k} \notin T$ we associate the cube $k_j \leqslant x_j \leqslant k_j + 1$, $j = 1, \cdots, s$. On this cube, $e^{-(\mathbf{h}, \mathbf{k})} \leqslant B_1 e^{-(\mathbf{h}, \mathbf{x})}$, where $B_1 = e^{h_1 + \cdots + h_s}$. Therefore, $e^{-(\mathbf{h}, \mathbf{k})}$ does not exceed the integral of the last function over the cube. It follows that the sum in (11) does not exceed the integral of $B_1 e^{-(\mathbf{h}, \mathbf{x})}$ over the region $x_j \geqslant 0$, $j = 1, \cdots, s$, $(\mathbf{h}, \mathbf{x}) \geqslant N$. Making the substitution $y_j = h_j x_j$, $j = 1, \cdots, s$, we obtain, by (7),

$$\sum_{\mathbf{k} \notin T}' e^{-(\mathbf{h}, \mathbf{k})} \leqslant \frac{B_1}{h_1 \cdots h_s} \int \cdots \int_{\Delta(N, \infty)} e^{-(y_1 + \cdots + y_s)} \, dy_1 \cdots dy_s$$

$$= B_2 \int_N^\infty e^{-y} y^{s-1} dy \leqslant B_3 N^{s-1} e^{-N}, \tag{12}$$

for some constants B_2, B_3. If N is given by (8), then, for all sufficiently small $\epsilon > 0$, $N^{s-1} \leqslant \text{const} \log^{s-1}(1/\epsilon)$. For sufficiently large L, we have from (11) and (12):

$$\| f - s_N(f) \| \leqslant \text{const} \, e^{-L} \epsilon \leqslant \tfrac{1}{2} \epsilon, \qquad f \in A.$$

We shall soon need estimates for the number P_N of the elements $\mathbf{k}$ of $T(N)$ and for the sum (14). When $\mathbf{k}$ runs through $T(N)$, the union of all cubes $k_j \leqslant x_j \leqslant k_j + 1$, $j = 1, \cdots, s$ is contained in the region $(\mathbf{h}, \mathbf{x}) \leqslant N + \alpha$, $\alpha = h_1 + \cdots + h_s$, $\mathbf{x} \geqslant 0$, and contains the region $(\mathbf{h}, \mathbf{x}) \leqslant N$, $\mathbf{x} \geqslant 0$. Therefore,

$$\int \cdots \int_{\substack{(\mathbf{h}, \mathbf{x}) \leqslant N \\ \mathbf{x} \geqslant 0}} dx_1 \cdots dx_s \leqslant P_N = \sum_{\mathbf{k} \in T(N)} 1 \leqslant \int \cdots \int_{\substack{(\mathbf{h}, \mathbf{x}) \leqslant N + \alpha \\ \mathbf{x} \geqslant 0}} dx_1 \cdots dx_s \,.$$

Using (6) we obtain

$$P_N = \frac{N^s}{s! h_1 \cdots h_s} \left(1 + O\left(\frac{1}{N}\right) \right). \tag{13}$$

In the same way,

$$\sum_{\mathbf{k} \in T} \log \mathbf{q}^{\mathbf{k}} = \sum_{\mathbf{k} \in T} (h_1 k_1 + \cdots + h_s k_s) = \frac{s N^{s+1}}{(s+1)! \, h_1 \cdots h_s} \left(1 + O\left(\frac{1}{N}\right) \right). \tag{14}$$

To construct the covering sets U, we put $\delta = \epsilon/(2P_N)$. Each set U will consist of functions $f \in A$, for which the Taylor coefficients $a_{\mathbf{k}}$, $\mathbf{k} \in T(N)$ are fixed, with errors not exceeding $\delta \mathbf{r}^{-\mathbf{k}}$. More formally, each U will be given by two sets $\{m_{\mathbf{k}}\}$, $\{m'_{\mathbf{k}}\}$, $\mathbf{k} \in T$, with integral entries $m_{\mathbf{k}}$, $m'_{\mathbf{k}}$. A function $f \in A$ belongs to U if and only if

$$\left.\begin{array}{l} m_{\mathbf{k}} \leqslant \dfrac{\mathbf{r^k}}{\delta} \operatorname{Re} a_{\mathbf{k}} < m_{\mathbf{k}} + 1 \\[2mm] m'_{\mathbf{k}} \leqslant \dfrac{\mathbf{r^k}}{\delta} \operatorname{Im} a_{\mathbf{k}} < m'_{\mathbf{k}} + 1 \end{array}\right\} \qquad \mathbf{k} \in T(N). \tag{15}$$

If two function f_1, f_2 belong to the same U, then by (2),

$$\| s_N(f_1) - s_N(f_2) \| \leqslant \sum_{\mathbf{k} \in T} 2\delta = 2\delta P_N = \epsilon,$$

and from (10) it follows that $\| f_1 - f_2 \| \leqslant 2\epsilon$, so that the sets U indeed form an ϵ-covering of A.

It remains to estimate the number of nonempty sets U. For $f \in A$ we have by (3), with $\mathbf{r'}$ instead of $\mathbf{r}$, $| a_{\mathbf{k}} | \mathbf{r^k} \leqslant Mq^{-\mathbf{k}}$. Hence, the number of values of $m_{\mathbf{k}}$, or of $m'_{\mathbf{k}}$, for which the corresponding interval (15) is not empty, is $\leqslant 2(Mq^{-\mathbf{k}}\delta^{-1} + 1)$, and the total number of nonempty sets U is at most

$$\prod_{\mathbf{k} \in T} 4 \left(\frac{2MP_N}{\epsilon \mathbf{q^k}} + 1 \right)^2. \tag{16}$$

Since $H_\epsilon(A)$ does not exceed the logarithm of (16), we obtain, using (13), (14), and (8),

$$H_\epsilon(A) \leqslant 2 \sum_{\mathbf{k} \in T} \left(\log \frac{1}{\epsilon} - \log \mathbf{q^k} + \log P_N + O(1) \right)$$

$$= 2P_N \log \frac{1}{\epsilon} - 2 \sum_{\mathbf{k} \in T} \log \mathbf{q^k} + O(P_N \log P_N)$$

$$= \frac{2}{(s+1)! \, h_1 \cdots h_s} \log^{s+1} \frac{1}{\epsilon} + O \left(\log^s \frac{1}{\epsilon} \log \log \frac{1}{\epsilon} \right). \tag{17}$$

(b) Taking an arbitrary $\epsilon > 0$, we shall find in A a large number of ϵ-distinguishable elements. From (2) we see that if

$$\sum_{\mathbf{k}} | a_{\mathbf{k}} | \mathbf{r^k} \leqslant M, \tag{18}$$

then the function $f(\mathbf{z}) = \sum a_{\mathbf{k}} \mathbf{z^k}$ belongs to A. Relation (18) can be assured by taking

$$| a_{\mathbf{k}} | \mathbf{r'^k} \leqslant \frac{M}{2^s} \frac{1}{\mathbf{k'^2}}, \qquad \mathbf{k'} = (k_1 + 1, \cdots, k_s + 1), \tag{19}$$

because

$$\sum \frac{1}{\mathbf{k'^2}} \leqslant \left(\sum_{k=1}^{\infty} \frac{1}{k^2} \right)^s \leqslant 2^s.$$

This time we define N by

$$N = \log \frac{1}{\epsilon} - 2s \log \log \frac{1}{\epsilon} - L, \tag{20}$$

where L is a constant to be determined later. As before, $T(N)$ is the set of all $\mathbf{k}$ for which $(\mathbf{h}, \mathbf{k}) \leqslant N$.

The functions f are defined by their Taylor coefficients. We put

$$
\begin{aligned}
a_\mathbf{k} &= 0 && \text{if} && \mathbf{k} \notin T, \\
a_\mathbf{k} &= (m_\mathbf{k} + i m'_\mathbf{k}) \, \mathbf{r}^{-\mathbf{k}} \sqrt{2}\epsilon && \text{if} && \mathbf{k} \in T,
\end{aligned}
\tag{21}
$$

where $m_\mathbf{k}$, $m'_\mathbf{k}$ are arbitrary integers that do not exceed, in absolute value $A_\mathbf{k} = M(2^{s+1} \epsilon \mathbf{k}'^2 \mathbf{q}^\mathbf{k})^{-1}$.

Then

$$| a_\mathbf{k} | \leqslant 2\epsilon \mathbf{r}^{-\mathbf{k}} A_\mathbf{k} = \frac{M}{2^s \mathbf{k}'^2 \mathbf{r}'^{\mathbf{k}}},$$

so that (19) is satisfied and all our functions f belong to A. Let f_1, f_2 be any two different functions of this type. Then at least one of the coefficients $a_\mathbf{k}$ of $f_1 - f_2$ is not zero. If we apply (3) to $f_1 - f_2$ and recall the definition of the norm, we obtain

$$\| f_1 - f_2 \| \geqslant | a_\mathbf{k} | \, \mathbf{r}^\mathbf{k} \geqslant \sqrt{2}\epsilon,$$

so that all functions f are 2ϵ-distinguishable.

It remains to count the number of different functions f. The bound $A_\mathbf{k}$ has been so selected that

$$A_\mathbf{k} \geqslant 1 \qquad \text{for} \qquad \mathbf{k} \in T. \tag{22}$$

In fact, $\mathbf{q}^{-\mathbf{k}} \geqslant e^{-N} = e^L \epsilon \log^{2s}(1/\epsilon)$, while $\mathbf{k}'^2 \leqslant \text{const } N^{2s}$, so that $A_\mathbf{k} \geqslant \text{const } e^L \geqslant 1$, if L is sufficiently large.

For a given $\mathbf{k} \in T$, the number of possible values of $m_\mathbf{k}$ or of $m'_\mathbf{k}$, is at least $2[A_\mathbf{k}] + 1$, and this is $\geqslant A_\mathbf{k}$ because of (22). Altogether, we obtain at least $\prod\limits_{\mathbf{k} \in T} A_\mathbf{k}^2$ different functions f. Therefore

$$C_\epsilon(A) \geqslant 2 \sum_{\mathbf{k} \in T} \log A_\mathbf{k} = 2 \sum_{\mathbf{k} \in T} \left\{ \log \frac{1}{\epsilon} - \log \mathbf{q}^\mathbf{k} - \log \mathbf{k}'^2 + O(1) \right\}.$$

But $\log \mathbf{k}'^2 = O(\log N) = O(\log P_N)$. Using the same computations as for (17), we see that

$$C_\epsilon(A) \geqslant \frac{2}{(s+1)! \, h_1 \cdots h_s} \log^{s+1} \frac{1}{\epsilon} + O\left(\log^s \frac{1}{\epsilon} \log \log \frac{1}{\epsilon} \right). \tag{23}$$

4. General Sets of Analytic Functions

Results similar to Theorems 4 and 5 hold for many other sets of functions $A^s(F, G; M)$. The proofs remain in principle the same, but instead of the Taylor series 3(1), one has to make use of other expansions adapted to the geometry of the pair of the regions F, G; for example, expansions into series of Chebyshev polynomials (if $F = [-1, +1]$ and $G = E_\rho$), series of Faber polynomials, and so on. The most general result for $s = 1$ (Erohin [46]) is that

$$C_\epsilon(A) \sim H_\epsilon(A) \sim \lambda \log^2 \frac{1}{\epsilon} \qquad (1)$$

holds for each set $A = A^1(F, G; M)$, where F is a closed, connected set, different from a point or the whole plane, and G is a "reasonable" region containing F. The constant λ depends upon some conformal invariants of the pair F, G. We shall prove only the following theorem for an arbitrary s.

THEOREM 6 (Kolmogorov [62]). If $A = A^s(F, G; M)$, if F has interior points, and G is bounded, then

$$C_\epsilon(A) \approx H_\epsilon(A) \approx \log^{s+1} \frac{1}{\epsilon} . \qquad (2)$$

Proof. We can find two concentric polycylinders, B: $|z_j - a_j| \leqslant r$, $j = 1, \cdots, s$, and B': $|z_j - a_j| < r'$, $j = 1, \cdots, s$, $0 < r < r'$, so that $B \subset F \subset G \subset B'$. Then $A' = A^s(B, B'; M)$, is contained in A as a point set and has smaller distances; therefore, $C_\epsilon(A) \geqslant C_\epsilon(A')$. By Theorem 5,

$$C_\epsilon(A) \geqslant a \log^{s+1} \frac{1}{\epsilon}, \qquad a > 0. \qquad (3)$$

It will be convenient to use

LEMMA 3. Let F be a compact Hausdorff space which is the union of its compact subsets F_k, $k = 1, \cdots, m$. Let $C[F]$ be the space of all complex continuous functions on F and let A be a compact subset of $C[F]$. If A_k, $k = 1, \cdots, m$ denote the sets of restrictions of all functions from A to F_k, $k = 1, \cdots, m$, with the distance derived from the uniform norm on F_k, then

$$H_\epsilon(A) \approx \max_k H_\epsilon(A_k). \qquad (4)$$

Proof. Each set A_k can be covered by $n_k = N_\epsilon(A_k)$ sets U_{lk}, $l = 1, \cdots, n_k$ of diameters $\leqslant 2\epsilon$. We define sets U in the following way: Each U is determined by a selection of numbers l_k, $1 \leqslant l_k \leqslant n_k$, $k = 1, \cdots, m$. A function $f \in A$ belongs to U if and only if, for each k, its restriction to F_k belongs to $U_{l_k k}$. The sets U form an ϵ-covering of A, and their number is $\prod_1^m n_k$. This implies $N_\epsilon(A) \leqslant \prod_1^m N_\epsilon(A_k)$. On the other hand, we have $N_\epsilon(A_k) \leqslant N_\epsilon(A)$. ∎

To complete the proof, we cover F by balls B_k of radii $r_k > 0$, $k = 1, \cdots, m$, and find concentric polycylinders B'_k with radii $r'_k > r_k$ and $B_k \subset B'_k \subset G$. Applying Lemma 3, we denote by A_k the set of the restrictions of all functions $f \in A$ to the set $F_k = F \cap B_k$. We compare the two sets: A_k and $A^s(B_k, B'_k; M)$. The second set contains the first (more exactly: each function of the first set has an extension that belongs to the second), and the distances in the second set are larger. By (4) and Theorem 5,

$$H_\epsilon(A) \leqslant \text{const} \max_k H_\epsilon(A^s(B_k, B'_k; M)) \leqslant b \log^{s+1} \frac{1}{\epsilon} \tag{5}$$

for some $b > 0$.   ∎

Notions of entropy can be applied in the theory of linear topological spaces. (We need only the basic notions of this theory, formulated below.) An example of a space of this type is the space $A^s(G)$ of functions analytic in the region G of the s-dimensional complex space. Its topology is generated by the following basis of neighborhoods $\mathfrak{U} = \{U\}$ of the origin: each $U = U(F, b)$ is defined by a compact subset F of G and a number $b > 0$; and U consists of all $f \in A^s(G)$ that satisfy $|f(z)| \leqslant b$ for all $z \in F$. This defines in $A^s(G)$ a linear topology (the topology of uniform convergence on compact sets).

A subset V of a linear topological space X is *bounded* in X if, for each neighborhood of the origin U, there is an $a > 0$ for which $V \subset aU$. Clearly, we may assume that U is an element of a basis $\mathfrak{U}$. It follows from the definitions, that a subset V of $A^s(G)$ is bounded if and only if the functions f of V are uniformly bounded on each compact set $F \subset G$.

We shall define the *approximate dimension* of a linear topological space X. For each neighborhood U of the origin in X, we define (compare the function $P_\epsilon(A)$ of Sec. 1) the number $P_\epsilon(V, U)$, which depends on $V \subset X$ and on $\epsilon > 0$. This is the smallest integer p such that p translations of ϵU cover V: $V \subset \bigcup_{k=1}^{p} (x_k + \epsilon U)$; if no such p exists, we put $P_\epsilon(V, U) = \infty$.

DEFINITION[1].   The approximate dimension of a linear topological space X is the class $\Phi(X)$ of all positive functions $\phi(\epsilon)$, defined for $\epsilon > 0$, that has the following property: For each bounded set $V \subset X$ and each neighborhood of the origin U, there is an $\epsilon_0 > 0$ for which

$$P_\epsilon(V, U) \leqslant \phi(\epsilon), \qquad 0 < \epsilon \leqslant \epsilon_0. \tag{6}$$

We must consider the space X large if the class $\Phi(X)$ is small. All notions involved in this definition are preserved under isomorphisms (that is, homeomorphisms that are also algebraic isomorphisms) of the linear topological

---

[1] There is a variation of this definition, in which *bounded* sets $V \subset X$ are replaced by *compact* sets $V \subset X$. For spaces $A^s(G)$, both definitions coincide: by a theorem of Montel, compact and closed bounded subsets of $A^s(G)$ are identical.

space X. It follows that the approximate dimension is *invariant under isomorphisms*.

As an example, we shall determine the approximate dimension of the s-dimensional euclidean space R_s. For each bounded subset V of R_s, and each neighborhood of the origin U, $P_\epsilon(V, U) \leqslant$ const ϵ^{-s} (see 9.1(9)), and there are V and U for which $P_\epsilon(V, U) \geqslant C\epsilon^{-s}$ for arbitrarily large C. Hence, $\Phi(R_s)$ consists of all functions ϕ for which $\lim_{\epsilon \to 0} \epsilon^s \phi(\epsilon) = +\infty$.

In a similar fashion we discuss the properties of functions $\phi \in \Phi(X)$ when $X = A^s(G)$. If a bounded set $V \subset X$ and a neighborhood $U = U(F, b) \in \mathfrak{U}$ are given, then (6) means there exist $p \leqslant \phi(\epsilon)$ functions $f_k \in A^s(G)$, $k = 1, \cdots, p$, for which the inequality $|f(z) - f_k(z)| \leqslant \epsilon b$, $z \in F$ holds for each $f \in V$ and some k. Putting $a = b^{-1}$, we obtain in terms of the entropy of V (with the metric derived from the uniform norm on F): The class $\Phi(A^s(G))$ consists of exactly those functions ϕ for which, for each selection of V, F and $a > 0$, there is a number $\epsilon_0 > 0$ such that

$$H_\epsilon(V) \leqslant \log \phi(a\epsilon), \qquad 0 < \epsilon \leqslant \epsilon_0. \tag{7}$$

As an example, we derive: If

$$\frac{\log \phi(\epsilon)}{\log^{s+1}(1/\epsilon)} \to \infty, \qquad \epsilon \to 0, \tag{8}$$

then $\phi \in \Phi(A^s(G))$. Indeed, if a bounded $V \subset A^s(G)$ and a $U = U(F, b)$ are given, we can find an open set G_1 with $F \subset G_1$, $\bar{G}_1 \subset G$. The functions $f \in V$ are bounded on $\bar{G}_1$, let us say by M. Then $V \subset A^s(F, G_1; M)$, and hence by Theorem 6,

$$H_\epsilon(V) \leqslant \text{const } \log^{s+1} \frac{1}{\epsilon} < \log \phi(a\epsilon)$$

for all small $\epsilon > 0$.

On the other hand, if $\log \phi(\epsilon)/\log^{s+1}(1/\epsilon) \leqslant c$ for $0 < \epsilon \leqslant \epsilon_0$, where $c > 0$ is sufficiently small, then $\phi \notin \Phi(A^s(G))$. For the proof, we take two concentric B, B' with $\bar{B} \subset G \subset B'$ and apply Theorem 6 to $V = A^s(B, B'; 1)$, which is a bounded set in $A^s(G)$.

As an interesting corollary of our remarks we have

THEOREM 7 (Kolmogorov [64]).   Two spaces $A^{s_1}(G_1)$ and $A^{s_2}(G^1)$ with $s_1 \neq s_2$ are not isomorphic.

5. Relations between Entropy and Widths

Both functions $H_\epsilon(A)$ and $d_n(A)$ measure the size of the set A, but they achieve this in different ways. It is not astonishing, therefore, that there exist relations between these functions, expressed by inequalities, and that these inequalities are not very strict.

Let A be a compact subset of a real Banach space X. In connection with the sequence $d_n = d_n(A)$, we define the following functions:

$$m(t) = \max_{n=1,2,\cdots} \{n : d_{n-1} \geqslant t^{-1}\}, \qquad t > 0, \tag{1}$$

$$l(t) = \max_{n=1,2,\cdots} \left\{n : \frac{d_{n-1}}{n} > t^{-1}\right\}, \qquad t > 0, \tag{2}$$

(the functions are zero if there are no such n).

THEOREM 8 (Mitjagin [80]). If A is a convex compact set with central symmetry (that is, $x \in A$ implies $-x \in A$), then

$$\int_0^{1/2\epsilon} \frac{l(t)}{t} \, dt \leqslant C_{2\epsilon}(A) \leqslant H_\epsilon(A) \leqslant m\left(\frac{2}{\epsilon}\right) \log \frac{4(d_0 + \epsilon)}{\epsilon}. \tag{3}$$

Proof. (a) Let $m = m(2/\epsilon)$. By (1), $d_m(A) < \frac{1}{2}\epsilon$. Let E_m be an m-dimensional subspace of X from which A deviates by at most $\frac{1}{2}\epsilon$. For each $x \in A$ there is a point $y \in E_m$ with $\rho(x, y) \leqslant \frac{1}{2}\epsilon$. Let A_m be the set of all points $y \in E_m$ with the property $\rho(A, y) \leqslant \frac{1}{2}\epsilon$. Each $\frac{1}{2}\epsilon$-net for A_m is also an ϵ-net for the set A. Therefore, $P_\epsilon(A) \leqslant P_{\epsilon/2}(A_m)$. By 9.1(3),

$$H_\epsilon(A) \leqslant \log M_{\epsilon/2}(A_m). \tag{4}$$

There exist $M = M_{\epsilon/2}(A)$ points $x_1, \cdots, x_M$ of A_m that are $\frac{1}{2}\epsilon$-distinguishable We consider the disjoint balls B_k, $k = 1, \cdots, M$ of the space E_m with centers x_k and radii $\frac{1}{4}\epsilon$. Since $\| x \| \leqslant d_0$ for $x \in A$, all balls B_k are contained in the ball B: $\| x \| \leqslant d_0 + \frac{1}{2}\epsilon$. To the space E_m, we assign an m-dimensional (euclidean) volume. For this purpose we select a basis $y_1, \cdots, y_n$ in E_m and map each point $y = \sum_{i=1}^m b_i y_i$ onto the point $(b_1, \cdots, b_m)$ of R_m. To a set in E_m we assign the volume equal to the volume of its image in R_m. The volume of a ball of radius r in E_m is λr^m, where λ does not depend on r. Comparing the volumes of the balls, we see that

$$M\left(\frac{\epsilon}{4}\right)^m \leqslant (d_0 + \epsilon)^m.$$

Taking logarithms, we obtain the right-hand inequality in (3).

(b) We first select a sequence of points $x_k \in A$. We take $x_1 \in A$ with $a_1 = \| x_1 \| = d_0$. If $x_1, \cdots, x_k$ are already known, we consider the linear space E_k that is spanned by these points. Let x_{k+1} be a point of A at the maximal distance from E_k. Then $a_{k+1} = \rho(x_{k+1}, E_k) \geqslant d_k$. We put $y_k = x_k/a_k$. The points y_k span the same spaces E_k, and

$$\rho(y_{k+1}, E_k) = 1, \qquad k = 0, 1, \cdots. \tag{5}$$

Because A is convex and symmetric with respect to the origin, for each n it contains, together with the points $a_k y_k = x_k$, the octahedron

$$\Delta_n = \left\{ y : y = \sum_1^n \lambda_k a_k y_k, \; \sum_1^n |\lambda_k| \leqslant 1 \right\}.$$

Clearly, $C_\epsilon(A) \geqslant C_\epsilon(\Delta_n)$.

In Δ_n we find many points that are ϵ-distinguishable. Let $\epsilon' > \epsilon$; we consider the points

$$y = \sum_1^n m_k \epsilon' y_k, \tag{6}$$

where the m_k are integers such that the corresponding y belong to Δ_n. Two different points y', y'' of this type are ϵ-distinguishable. In fact, if k is the largest index for which they have different coefficients in (6), then by (5),

$$\| y' - y'' \| \geqslant \rho(\epsilon' y_k, E_{k-1}) = \epsilon' > \epsilon.$$

We estimate the number N of different points y. For this purpose we assign to the space E_n the n-dimensional volume, which corresponds to the mapping $\sum_{k=1}^n b_k y_k \rightarrow (b_1, \cdots, b_n)$ of E_n onto R_n. The octahedron Δ_n has the volume $2^n(1/n!)\, a_1 \cdots a_n$. The n-dimensional cubes $S(y)$ with centers y and sides $2\epsilon'$ have volumes $(2\epsilon')^n$. They cover Δ_n; hence,

$$\frac{1}{n!} a_1 \cdots a_n \leqslant N \epsilon'^n.$$

From this we obtain

$$C_\epsilon(\Delta_n) \geqslant \log N \geqslant \sum_{k=1}^n \log \frac{a_k}{k \epsilon'}. \tag{7}$$

We can replace a_k here by d_{k-1} and ϵ' by ϵ. We select n so that the resulting sum in (7) becomes as large as possible:

$$C_\epsilon(A) \geqslant \sum_{d_{k-1}/(k\epsilon) \geqslant 1} \log \frac{d_{k-1}}{k \epsilon}.$$

The function $l(t)$ has jumps equal to 1 at each of the points $t = k/d_{k-1}$, $k = 1, \cdots, n$, and vanishes for $0 \leqslant t \leqslant d_0^{-1}$. Therefore, the last sum is equal to

$$\int_{1/d_0}^{1/\epsilon} \log \frac{1}{\epsilon t}\, dl(t) = l(t) \log \frac{1}{\epsilon t} \Big|_{1/d_0}^{1/\epsilon} + \int_{1/d_0}^{1/\epsilon} \frac{l(t)}{t}\, dt = \int_0^{1/\epsilon} \frac{l(t)}{t}\, dt. \qquad \blacksquare$$

6. Notes

1. It seems to be very difficult to improve the results of Theorem 3, 4, and 5, and to obtain sharper estimates. Among the few known results we mention the formula [65]:

$$H_\epsilon(A) = \frac{lM}{\epsilon} + \log \frac{C}{\epsilon} + O(1). \tag{1}$$

Here, A is the class of functions that belong to $\mathrm{Lip}_M 1$ on an interval of length l and are bounded by C in absolute value. See also Tihomirov [94].

2. Entropies of many other sets, in particular of analytic functions, are computed in [65]. See also Clements [39].

3. Vituškin [17] has successfully applied entropy to obtain lower bounds of the degree of nonlinear approximation, for classes of differentiable and analytic functions. At present, this is the only known general approach to this problem. Very roughly, Vituškin's results show that the degree of approximation of large classes of functions cannot improve appreciably if the approximating polynomials $a_1 \phi_1(x) + \cdots + a_n \phi(x)$ are replaced by a family of functions $F(a_1, \cdots, a_n; x)$, which depend in a simple nonlinear way upon n parameters, $a_1, \cdots, a_n$.

4. Approximate dimension and entropy are important tools for the classification of linear topological spaces. In particular, *nuclear spaces* have been characterized (Mitjagin [80]) among all F-spaces X by the relation

$$\frac{\log \log P_\epsilon(A, V)}{\log (1/\epsilon)} \to 0, \qquad \epsilon \to 0, \tag{2}$$

which must hold for each compact set $A \subset X$ and each neighborhood U of the origin.

5. For which sets A can the entropy $H_\epsilon(A)$ be determined, if the approximation properties of A are known? This is possible for each full approximation set $A = A(\Delta) = A(\Phi, \Delta)$ in an arbitrary Banach space (see p. 139). Let $\delta_n > 0$ be a sequence that decreases to zero. We put

$$N_i = \min \{k : \delta_k \leqslant e^{-i}\}, \qquad i = 1, 2, \cdots$$

and define n by the inequality $e^{-n+1} \leqslant \epsilon < e^{-n+2}$. Then (Lorentz [77])

$$N_1 + \cdots + N_{n-3} \leqslant H_\epsilon(A) \leqslant N_1 + \cdots + N_n + 2N_n$$
$$+ \sum_{i=1}^{n-1} \Delta N_i \log \frac{N_n}{\Delta N_i} + Q(1). \tag{3}$$

For the special cases of the sets $\Lambda_{p\omega}^s$ and $A^s(\mathbf{r}, \mathbf{r}'; M)$, one obtains exactly the estimates 2(2) and 3(5) of the present chapter. For the first set, the necessary information about its approximation properties is contained in Theorem 8, Chapter 6; for the second set, it is derived from the approximation of $f \in A^s(\mathbf{r}, \mathbf{r}'; M)$ by the partial sums of their Taylor series. Similar results hold in spaces of functions with norms different from the uniform norm. Let $\text{Lip}(\alpha, p)$, $p > 1$, $0 < \alpha \leqslant 1$, be the subset of $L^p[a, b]$, that consists of all functions f for which $\{\int_a^b |f(x + t) - f(x)|^p \, dx\}^{1/p} \leqslant |t|^\alpha$. Then

$$H_\epsilon(\text{Lip}(\alpha, p)) \approx (1/\epsilon)^{1/\alpha}.$$

PROBLEMS

1. Let $A = \bigcup_{k=1}^s A_k$; then $N_\epsilon(A) \approx \max_k N(A_k)$.

2. A metric space is centerable if each of its subsets is centerable. Prove that the euclidean plane is not centerable.

3. Prove that each compact metric space A can be isometrically imbedded into the centerable Banach space m, which consists of all bounded real sequences $y = \{y_n\}$ with $\|y\| = \sup |y_n|$. [*Hint:* Let x_n be dense in A; put $y = \{\rho(x, x_n)\}$.]

4. Prove that $H_\epsilon(A) = \min H_\epsilon^X(A)$, where the minimum is taken for all metric spaces X that contain A as a subspace.

5. If A is an infinite dimensional, convex set in a Banach space X, then

$$\lim_{\epsilon \to \infty} \left[\frac{C_\epsilon(A)}{\log (1/\epsilon)}\right] = \infty.$$

6. Let X be an infinite dimensional Banach space, and $\phi(\epsilon)$ a decreasing positive function, defined for $\epsilon > 0$. There exists a compact subset A of X with $M_\epsilon(A) \geqslant \phi(\epsilon)$.

7. Find the approximative dimension $\Phi(X)$ of the following spaces X: an infinitely-dimensional Banach space; the space $A^s(D)$, where D is a polycylinder.

8. Find relations between the entropies and capacities of A and of Λ_ω, where A is a compact metric space, ω a modulus of continuity, and Λ_ω consists of all real functions f on A, which vanish at a given point x_0 and satisfy $\omega(f, h) \leqslant \omega(h)$.

9. Let $A'(\rho, C)$ be the set of all complex-valued functions on $[-1, +1]$, with uniform norm, having an analytic extension into the ellipse $E\rho$, which is bounded in absolute value by C. Prove that

$$H_\epsilon(A') \sim \frac{\log^2 (1/\epsilon)}{\log \rho}.$$

10. Answer the same question for functions with *real* values on $[-1, +1]$.

[11]

Representation of Functions of Several Variables by Functions of One Variable

1. The Theorem of Kolmogorov

In this chapter we shall discuss the representation of continuous functions of several variables by superpositions of functions of one variable and by sums of functions. Even a beginning student may notice that examples of genuine functions of two variables are rare in a course of elementary Calculus. Of course $x + y$ is one such function. But all other functions known to him reduce to this trivial one and to functions of one variable; for example, $xy = e^{\log x + \log y}$.

This serves as an illustration of the following theorem of Kolmogorov. Let $I = [0, 1]$ and let S be the s-dimensional cube, $0 \leqslant x_p \leqslant 1$, $p = 1, \cdots, s$.

THEOREM 1. There exist s constants $0 < \lambda_p \leqslant 1$, $p = 1, \cdots, s$, and $2s + 1$ functions $\phi_q(x)$, $q = 0, \cdots, 2s$, defined on I and with values in I, which have the following properties: The ϕ_q are strictly increasing and belong to a class Lip α, $\alpha > 0$. For each continuous function f on S, one can find a continuous function $g(u)$, $0 \leqslant u \leqslant s$ such that

$$f(x_1, \cdots, x_s) = \sum_{q=0}^{2s} g(\lambda_1 \phi_q(x_1) + \cdots + \lambda_s \phi_q(x_s)). \tag{1}$$

It will be seen that for $s = 2$, we can take $\lambda_1 = 1$, $\lambda_2 = \lambda$. We have, therefore, for each continuous function $f(x, y)$, $0 \leqslant x, y \leqslant 1$ a representation

$$f(x, y) = \sum_{q=0}^{4} g(\phi_q(x) + \lambda \phi_q(y)). \tag{2}$$

Formulas (1) and (2) reduce the function f to sums and superpositions (that is, functions of functions) of functions of one variable, g, $\lambda_p \phi_q$. The function g depends on f, but the $(2s + 1) s$ functions $\lambda_p \phi_q$ do not.

The history of this theorem is very interesting. In his famous lecture at the International Conference of Mathematicians in Paris, 1900, Hilbert formulated 23 problems, which in his opinion were important for the further development of mathematics. They have since attracted the attention of many outstanding mathematicians. The thirteenth of these problems contained (implicitly) the conjecture that not all continuous functions of three variables are representable as superpositions of continuous functions of two variables. This conjecture was refuted in 1957 by Kolmogorov and his pupil, Arnol'd. In the last of their three papers on this subject, Kolmogorov [63] proved Theorem 1. It is clear that Theorem 1 refutes Hilbert's conjecture, since sums of several terms that appear in the formula (1) can be built up of sums of two terms. Kolmogorov's original formulation had the representation

$$f(x_1, \cdots, x_s) = \sum_{p=0}^{2s} g_p \left(\sum_{q=1}^{s} \phi_{pq}(x_p) \right) \qquad (3)$$

instead of (1). The possibility of replacing the functions g_p by a single function (without increasing the interval of the definition) has been noticed in [74], and Sprecher [89] proved that one can take $\phi_{pq} = \lambda_p \phi_q$. The proof below, outlined in [74], follows the ideas of the original proof of Kolmogorov [63].

In order to obtain a geometric interpretation of Theorem 1, let us consider the map of S into the $(2s + 1)$-dimensional space given by

$$z_p = \lambda_1 \phi_p(x_1) + \cdots + \lambda_s \phi_p(x_s), \qquad p = 0, 1, \cdots, 2s. \qquad (4)$$

This map is continuous. It is also one-to-one. For, otherwise, there would exist two points of S, which are not distinguished by the family of functions $\lambda_1 \phi_p(x_1) + \cdots + \lambda_s \phi_p(x_s)$, $p = 0, \cdots, 2s$. Then all functions representable by the sum (1) would coincide at these two points, and the representation (1) would be impossible for some functions $f \in C[S]$.

Since S is compact, its image T under (4) is compact, and (4) is a homeomorphism between S and T. It follows that there is a one-to-one correspondence between all continuous functions $f(x_1, \cdots, x_s)$ on S and all continuous functions $F(z_0, \cdots, z_{2s})$ on T. Therefore, Theorem 1 can be stated as follows: There exists a homeomorphic imbedding of S into the $(2s + 1)$-dimensional euclidean space, of the special form (4), so that each continuous function F on the image of S has the form

$$F(z_0, \cdots, z_{2s}) = \sum_{p=0}^{2s} g(z_p). \qquad (5)$$

The following two sections contain the proof of Theorem 1. For the constants λ_p of this theorem, one can take arbitrary rationally independent numbers in the range $0 < \lambda_p < 1$, that is, any numbers λ_p, for which a relation $r_1 \lambda_1 + \cdots + r_s \lambda_s = 0$, with rational r_p is possible only if $r_1 = \cdots = r_s = 0$. Accordingly, in case $s = 2$, we may assume that $\lambda_1 = 1$ and $\lambda_2 = \lambda$ is irrational.[1]

The proof of Theorem 1 will be restricted to the case $s = 2$. This will simplify the notation, but will have no influence on the ideas of the argument.

---

[1] Actually, these conditions can be relaxed; in particular, for $s = 2$, λ, with $0 < \lambda < 1$ may be taken arbitrarily. But we do not wish to insist here on details.

2. The Fundamental Lemma

In this section we shall construct the functions ϕ_q. We define the closed intervals

$$I_i^k = [i \cdot 10^{-k+1} + 10^{-k}, i \cdot 10^{-k+1} + 9 \cdot 10^{-k}], \quad i = 0, \cdots, 10^{k-1}, \quad k = 1, 2, \cdots; \tag{1}$$

k will be called the *rank* of this interval. All intervals I_i^k are contained in $I = [0, 1]$, except when $i = 10^{k-1}$.

To illustrate our idea, let us recall, [26, p. 213] the construction of an increasing function $\psi(x)$, $0 \leqslant x \leqslant 1$, which is not constant and has derivative 0 almost everywhere. We adapt this construction to the intervals (1).

We put $\psi(0) = 0$, $\psi(1) = 1$. Next we define $\psi(x) = \frac{1}{2}$ on $I_0^1 = [10^{-1}, 9 \cdot 10^{-1}]$, the only interval of rank 1 contained in I. At the next step, we define ψ on the intervals of rank 2 not contained in I_0^1, putting $\psi(x) = \frac{1}{4}$ for $x \in I_0^2$ and $\psi(x) = \frac{3}{4}$ for $x \in I_9^2$. This is continued indefinitely. In this way, ψ is defined on the union of all intervals (1), that is, on the set $A = \cup I_i^k$. It is easy to prove (see [26, p. 213]) that ψ has a continuous increasing extension onto I, given by $\psi(x) = \sup_{0 \leqslant y \leqslant x} \psi(y)$. The sum of the lengths of the intervals I_i^k that have been used in the construction is $8 \cdot 10^{-1}[1 + (\frac{2}{10}) + (\frac{2}{10})^2 + \cdots] = 1$, and it is clear that $\psi'(x) = 0$ inside each of the intervals. Hence, indeed $\psi'(x) = 0$ a.e. on I.

The functions ϕ_q are similar to ψ. The qth function ϕ_q, $q = 0, \cdots, 4$, is connected with the intervals

$$I_{qi}^k = [\alpha_{qi}^k, \beta_{qi}^k] = I_i^k - 2q \cdot 10^{-k},$$

$$k = 1, 2, \cdots; \qquad i = 0, 1, \cdots, 10^{k-1}, \tag{2}$$

which are obtained from the I_i^k by the translation to the left by the distance $2q10^{-k}$. (This distance depends on the rank k.) We replace by $I \cap I_{qi}^k$ those intervals I_{qi}^k that are not entirely contained in I.

The intervals of rank k are of length $8 \cdot 10^{-k}$, and the gaps between them are of length $2 \cdot 10^{-k}$. If we fix k, the intervals (2) form five families, corresponding to $q = 0, \cdots, 4$. No one of these families covers I. However, each point $x \in I$ is covered by at least *four* of the *five* families, $q = 0, \cdots, 4$.

Another useful remark is that the end points α_{qi}^l, β_{qi}^l of ranks $l < k$ are among the points of the form $m \cdot 10^{-k+1}$; and that no end points of rank k (except perhaps 0, 1) are of this form. Therefore, none of the α_{qi}^k, β_{qi}^k different from 0, 1 coincides with an end point of lower rank.

Unlike ψ, the functions ϕ_q will be strictly increasing. Their most important property will be that of mapping the intervals I_{qi}^k of each given rank into small sets. It is not the smallness of measure that is important for us, but rather the fact that the images of the I_{qi}^k under the ϕ_q will be disjoint, not only for a fixed

q—this would follow from the monotony of the ϕ_q—but for all q and all i (and a fixed rank k). Actually, even more will be needed.

The exact formulation of the required property is given in the following lemma. Let $0 < \lambda < 1$ be *irrational*. We note that the relation $r + r'\lambda = r_1 + r_1'\lambda$ with rational r's is possible only if $r = r_1$, $r' = r_1'$.

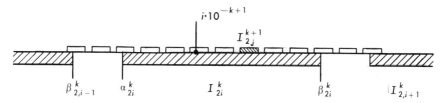

FIG. 2. Intervals I_{qi}^k for $q = 2$.

LEMMA 1. There exist five strictly increasing functions ϕ_q, $q = 0, \cdots, 4$, defined on $[0, 1]$, with values in $[0, 1]$, and belonging to a class Lip α, $\alpha > 0$ such that for each fixed rank $k = 1, 2, \cdots$, the intervals

$$\Delta_{qij}^k = [\phi_q(\alpha_{qi}^k) + \lambda\phi_q(\alpha_{qj}^k), \qquad \phi_q(\beta_{qi}^k) + \lambda\phi_q(\beta_{qj}^k)], \qquad q = 0, \cdots, 4,$$

$$i, j = 0, \cdots, 10^{k-1} \qquad (3)$$

are all disjoint.

Proof. The functions ϕ_q will be constructed simultaneously. We put $\alpha_{q0}^0 = 0$, $\beta_{q0}^0 = 1$, and define $0 \leqslant \phi_q(0) < \phi_q(1) < 1$, $q = 0, \cdots, 4$, in such a way that all these values are rational and distinct. We continue this construction by induction on k. At the kth step, we shall define each ϕ_q, $q = 0, \cdots, 4$, at the end points α_{qi}^k, β_{qi}^k of all intervals (2) with the same q. For each $k = 1, 2, \cdots$, we shall take care to satisfy the following conditions:

(a) The function ϕ_q, $q = 0, \cdots, 4$, is strictly increasing on the set of all points α_{qi}^l, β_{qi}^l, $0 \leqslant i \leqslant 10^{l-1}$, $l \leqslant k$.

(b) The slope of each segment of the polygonal line connecting the points $(x, \phi_q(x))$, $x = \alpha_{qi}^l$, β_{qi}^l, $l \leqslant k$, is strictly smaller than 5^k.

The preceding properties involve each single function ϕ_q; the following properties involve all five functions.

(c) All values $\phi_q(\alpha_{qi}^l)$, $\phi_q(\beta_{qi}^l)$, $q = 0, \cdots, 4$, $0 \leqslant i \leqslant 10^{k-1}$, $l \leqslant k$, are rational and distinct.

(d) The intervals (3) are disjoint.

Assume that the ϕ_q are defined at all end points α_{qi}^l, β_{qi}^l of ranks $l < k$ in such a way that conditions (a) to (d) are satisfied. We shall explain how this definition can be extended to all end points α_{qi}^k, β_{qi}^k in $(0, 1)$ of rank k. First we construct a (not strictly) increasing extension of ϕ_q that satisfies (b). Later we shall amend this extension slightly so as to satisfy all the requirements (a) to (d). The first part of the construction can be carried out independently for each q.

Let q be fixed. We shall denote by $\gamma = \gamma_q$ the points α_{qi}^l, β_{qi}^l, $l < k$, where the function ϕ_q is already known. We have observed (on p. 170) that each "new" point α_{qi}^k, β_{qi}^k in $(0, 1)$ is different from each of the old points γ. The intervals I_{qi}^k cover $\frac{4}{5}$ of the total length of each interval $(m10^{-k+1}, (m + 1) 10^{-k+1})$, and the gaps between the intervals I_{qi}^k cover the remaining $\frac{1}{5}$ of this length. The same applies to each interval (γ, γ'). The intervals I_{qi}^k are of two kinds. The intervals of the first kind contain one point γ; we call it γ_{qi}^k, the intervals of the second kind contains no points γ.

For each interval I_{qi}^k, we shall take $\phi_q(\alpha_{qi}^k) = \phi_q(\beta_{qi}^k) = \phi_{qi}^k$. For the intervals of the first kind, we select ϕ_{qi}^k equal to $\phi_q(\gamma_{qi}^k)$. To define ϕ_{qi}^k for the intervals I_{qi}^k of the second kind, we select the values of ϕ_q at the end points α_{qi}^k, β_{qi}^k. Let γ, γ' be two adjacent points γ_q. We define ϕ_q at the end points of the intervals I_{qi}^k of the second kind contained in (γ, γ'). It is sufficient to determine the increase of ϕ_q on each gap (or a part of a gap) (x, x') between two intervals I_{qi}^k, which falls into (γ, γ'). By (b), $\phi_q(\gamma') - \phi_q(\gamma) < 5^{k-1}(\gamma' - \gamma)$. To (x, x') we assign an increase of ϕ_q, proportional to the length of (x, x'). Since the total length of the gaps is $\frac{1}{5} (\gamma', \gamma)$, we must take

$$\phi_q(x') - \phi_q(x) = 5 \frac{x' - x}{\gamma' - \gamma} [\phi_q(\gamma') - \phi_q(\gamma)].$$

Thus, $\phi_q(x') - \phi_q(x) < 5^k(x' - x)$, and ϕ_q satisfies (b).

From now on the construction involves all values of q. Small changes of the values of the ϕ_{qi}^k will not disturb the condition (b), or the monotonicity of ϕ_q. We change slightly the ϕ_{qi}^k for the intervals of the second kind in such a way that all values ϕ_{qi}^k, $q = 0, \cdots, 4$, $i = 0, \cdots, 10^{k-1}$, become rational and distinct. This is possible because, in view of (c), the values $\phi_{qi}^k = \phi_q(\gamma_{qi}^k)$ for the intervals of the first kind already have these properties.

The last part of the kth step is an amendment of ϕ_q at the end points α_{qi}^k, β_{qi}^k of rank k, which lie in $(0, 1)$. Since λ is irrational, the values

$$\phi_{qi}^k + \lambda\phi_{qj}^k, \qquad q = 0, \cdots, 4; \qquad i, j = 0, \cdots, 10^{k-1} \qquad (4)$$

are distinct. Let $\epsilon > 0$ be so small that the 2ϵ-neighborhoods of the points (4) are disjoint. This allows us to define the remaining values of the ϕ_q. For each interval I_{qi}^k, we select $\phi_q(\alpha_{qi}^k)$, $\phi_q(\beta_{qi}^k)$ in the ϵ-neighborhood of ϕ_{qi}^k in such a way that $\phi_q(\alpha_{qi}^k) < \phi_{qi}^k < \phi_q(\beta_{qi}^k)$. (If one of the end points is 0 or 1, this inequality has to be changed correspondingly; for example, if $\beta_{qi}^k = 1$, it should read $\phi_q(\alpha_{qi}^k) < \phi_{qi}^k = \phi_q(\beta_{qi}^k)$.) It is clear that these selections can be made so that (a), (b), and (c) will be satisfied; (d) will also hold because each interval (3) is contained in the 2ϵ-neighborhood of the corresponding point (4). This completes the kth step.

By induction, each function ϕ_q is defined on the set A_q of the points α_{qi}^k, β_{qi}^k, $0 \leqslant i \leqslant 10^{k-1}$, $k = 1, 2, \cdots$. Moreover, on A_q,

$$\phi_q \in \text{Lip } \alpha, \qquad \alpha = \log_{10} 2. \tag{5}$$

Indeed, let x, $x + h$, $h > 0$ be two points of A_q. We may assume $h < \frac{1}{5}$. Let k be the integer that satisfies $2 \cdot 10^{-k-1} \leqslant h < 2 \cdot 10^{-k}$. The interval $(x, x + h)$ can contain at most one of the points α_{qi}^k, β_{qi}^k. Hence, it is contained in an interval of length 10^{-k+1} between some two such points. Therefore by (a) and (b),

$$\phi_q(x + h) - \phi_q(x) < 5^k \cdot 10^{-k+1} = 10 \cdot 2^{-k} = 10 \cdot 10^{-\alpha k} \leqslant 10(5h)^\alpha = 10 \cdot 5^\alpha h^\alpha$$

if α is defined by $2 = 10^\alpha$. This proves (5).

The set A_q is dense in $[0, 1]$. By uniform continuity, ϕ_q can be uniquely extended onto $[0, 1]$. This extension $\phi_q(x)$ will still be strictly increasing and will satisfy (5). The extension will not change the values of ϕ_q at the α_q, β_q, so that the new functions will also satisfy (d). ∎

3. The Completion of the Proof

Let I_{qi}^k be the intervals 2(2); let S be the square $0 \leqslant x, y \leqslant 1$; and let S_{qij}^k be the little squares:

$$S_{qij}^k = I_{qi}^k \times I_{qj}^k, \qquad q = 0, \cdots, 4; \qquad k = 1, 2, \cdots; \qquad 0 \leqslant i, j \leqslant 10^{k-1}. \tag{1}$$

The image of a square (1) under the map $(x, y) \to \phi_q(x) + \lambda\phi_q(y)$ is the interval Δ_{qij}^k of 2(3). Lemma 1 means, therefore, that for each fixed k, the functions $\phi_q(x) + \lambda\phi_q(y)$ map squares (1) with the corresponding q onto disjoint intervals.

For fixed k, the intervals I_{qi}^k form five families, $q = 0, \cdots, 4$. We know that each point x in $[0, 1]$ can fail to be covered only by one of the five families. The same applies to the intervals I_{qj}^k. Therefore, each point (x, y) of S can fail to be covered by the squares (1) only for at most two values of q. We have "at least three hits at (x, y) out of five tries, $q = 0, \cdots, 4$."

For each continuous function $g(u)$, defined for $0 \leqslant u \leqslant 2$, we put

$$h(x, y) = \sum_{q=0}^{4} g(\phi_q(x) + \lambda\phi_q(y)). \tag{2}$$

LEMMA 2. Let $\frac{2}{3} < \theta < 1$. For each $f \in C[S]$ there is a function $g \in C[0, 2]$ such that

$$\|f - h\| \leqslant \theta \|f\|; \qquad \|g\| \leqslant \tfrac{1}{3}\|f\|. \tag{3}$$

Proof. We take $\epsilon > 0$ so small that $\frac{2}{3} + \epsilon \leqslant \theta$, and then k so large that the oscillation of f on each of the squares (1) does not exceed $\epsilon \|f\|$. Let f_{qij}^k be the value of $f(x, y)$ in the center of the square S_{qij}^k. On each interval Δ_{qij}^k we take $g(u)$ constant and equal to $\frac{1}{3} f_{qij}^k$. We can extend g linearly into the gaps between the Δ_{qij}^k, and obtain in this way a continuous function on $[0, 2]$ that satisfies the second relation (3). Let (x, y) be an arbitrary point of S. For at least three values of q, $(x, y) \in S_{qij}^k$ for some i and j. The corresponding terms of the sum (2) have the values $\frac{1}{3} f_{qij}^k$, each of which is equal to $\frac{1}{3} f(x, y)$, with an error less than $\frac{1}{3} \epsilon \|f\|$. The two remaining terms of (2) are each $\leqslant \frac{1}{3} \|f\|$ in absolute value. It follows that

$$|f(x, y) - h(x, y)| \leqslant 3 \tfrac{1}{3} \epsilon \|f\| + \tfrac{2}{3} \|f\| \leqslant \theta \|f\|. \qquad \blacksquare$$

Proof of Theorem 1 (for $s = 2$).   Let $f \in C[S]$ be given. We define a sequence of functions $g_r \in C[0, 2]$ with the corresponding functions

$$h_r(x, y) = \sum_{q=0}^{4} g_r(\phi_q(x) + \lambda \phi_q(y))$$

as follows: First, let g_1, h_1 be given by Lemma 2 with

$$\|f - h_1\| \leqslant \theta \|f\|, \qquad \|g_1\| \leqslant \tfrac{1}{3} \|f\|.$$

We apply Lemma 2 again and obtain functions g_2, h_2 for which

$$\|(f - h_1) - h_2\| \leqslant \theta \|f - h_1\| \leqslant \theta^2 \|f\|, \qquad \|g_2\| \leqslant \tfrac{1}{3} \|f - h_1\| \leqslant \tfrac{1}{3} \theta \|f\|.$$

In general, the functions g_r, h_r will satisfy

$$\|f - h_1 - \cdots - h_r\| \leqslant \theta^r \|f\|,$$

$$\|g_r\| \leqslant \tfrac{1}{3} \theta^{r-1} \|f\|, \qquad r = 1, 2, \cdots. \tag{4}$$

The series $\sum_{1}^{\infty} g_r$, $\sum_{1}^{\infty} h_r$ converge uniformly. Let g, h be their sums. They are connected by the relation (2). Moreover, from the first relation (4), $f = h$. This gives us the desired representation 1(2). $\blacksquare$

4. Functions Not Representable by Superpositions

Although Hilbert's conjecture has been disproved by Theorem 1, it was based on the sound idea that "bad" functions cannot be represented in a simple way by "good" functions. Kolmogorov's theorem shows only that, for our purposes, the number of variables s is not a satisfactory characteristic of "badness" of a continuous function. But for functions that have continuous derivatives of order $p \geqslant 1$, the number $\chi = p$ is a useful characteristic. Indeed, it is a simple fact of the differential calculus that not all functions with continuous partial derivatives of order $p - 1$ can be represented by superpositions of functions with continuous derivatives of order p.

The purpose of this section is to show that there exist characteristics χ, with similar properties that *depend* on s. For the classes $\Lambda^s_{p\alpha}$ (see Chapter 3, Sec. 7) of functions defined on an s-dimensional unit cube B, a characteristic of this type is the number $\chi = (p + \alpha)/s$. For the more general classes $\Lambda^s_{p\omega}$, this χ is replaced by the function $\Delta(\epsilon) = \delta(\epsilon)^s$, where $\delta(\epsilon)$ is given by 10.2(1). If $\omega(h) = Mh^\alpha$, then $\Delta(\epsilon) = \text{const } \epsilon^{1/\chi}$. Small χ, or functions $\Delta(\epsilon)$ that decrease rapidly to zero, correspond to "big," "complicated" classes Λ. We prove the following theorem, due to Vituškin [100] and Kolmogorov [65].

THEOREM 2.   Let $\chi_0 > 0$ be given. Let L be the union of all classes $\Lambda^s_{p\alpha}$ with $\chi = (p + \alpha)/s > \chi_0$ and $p + \alpha \geqslant 1$. Then not all functions of a class $\Lambda^{s_0}_{p_0\alpha_0}$, $(p_0 + \alpha_0)/s_0 = \chi_0$ can be represented as superpositions of functions of L.

Roughly, not all functions of a given characteristic χ_0 can be represented as superpositions of simpler functions. Somewhat more generally, we have

THEOREM 3.   Let F be a countable family of classes $\Lambda^s_{p\omega} (M_0, \cdots, M_{p+1}; B)$ (each of which is contained in a class $\text{Lip}_M 1$), with the corresponding functions $\Delta(\epsilon)$. Let a class $\Lambda^{s_0}_{p_0\omega_0}$ with the function $\Delta_0(\epsilon)$ be given. If

$$\lim_{\epsilon \to 0} \frac{\Delta_0(\epsilon)}{\Delta(C\epsilon)} = 0 \tag{1}$$

for each $C > 0$ and each Δ, then not all functions of the class $\Lambda^{s_0}_{p_0\omega_0}$ are superpositions of functions of $L = \bigcup_F \Lambda$.

In Theorem 2, the class L can be thought of as a *countable* union of classes $\Lambda^s_{p\alpha}$, namely, as the union of all classes $\Lambda^s_{p\alpha}(M_0, \cdots, M_{p+1}; B) \subset \text{Lip}_M 1$ for which $(p + \alpha)/s > \chi_0$, $p + \alpha \geqslant 1$, α is *rational*, and $M_0, \cdots, M_{p+1}$, M are positive *integers*. Therefore, Theorem 2 is a special case of Theorem 3.

Before we prove Theorem 3, we must introduce some new notions. They will help us to describe all possible superpositions. Consider, for example, the function $g(h(x, y), z)$. We can write this as $g(y_1, y_2)$, where $y_1 = h(y_{11}, y_{12}), y_2 = z$ and $y_{11} = x, y_{12} = y$. In this example, the basic variable z is reached in one step, while two steps are needed to reach the basic variables x and y. We can make the number of steps equal by introducing intermediate variables, $y_2 = y_{21}, y_{21} = z$. This leads to the following definitions. A *scheme* S is a table of natural numbers of the following type:

$$S: \begin{cases} s, n \\ m \\ m_{k_1}, \; k_1 = 1, \cdots, m \\ m_{k_1 k_2}, \; k_1 = 1, \cdots, m; \; k_2 = 1, \cdots, m_{k_1} \\ \cdots \cdots \cdots \cdots \cdots \cdots \cdots \cdots \cdots \\ m_{k_1 \ldots k_n}, \; k_1 = 1, \cdots, m; \; \cdots; \; k_n = 1, \cdots, m_{k_1 \ldots k_{n-1}} \\ \text{each } m_{k_1 \ldots k_n} \text{ is one of the numbers } 1, \cdots, s. \end{cases}$$

With S we associate *admissible sets of subscripts*. These are all sets of natural numbers $k_1, \cdots, k_r$, $0 \leqslant r < s$, which appear as subscripts of the integers m in S. The value $r = 0$ is not excluded and gives the empty set of subscripts.

If a scheme S is given, with each admissible set of subscripts we associate a function $g_{k_1 \ldots k_r}$ of $m_{k_1 \ldots k_r}$ variables. We set up the formulas:

$$
\left\{
\begin{aligned}
& f(x_1, \cdots, x_s) = g(y_1, \cdots, y_m); \\
& \qquad y_{k_1} = g_{k_1}(y_{k_1 1}, \cdots, y_{k_1 m_{k_1}}), \\
& \qquad \cdots \cdots \cdots \cdots \cdots \cdots \cdots \cdots \cdots \\
& \quad y_{k_1 \ldots k_{n-1}} = g_{k_1 \ldots k_{n-1}}(y_{k_1 \ldots k_{n-1} 1}, \cdots, y_{k_1 \ldots k_{n-1} m_{k_1 \ldots k_{n-1}}}), \\
& \qquad y_{k_1 \ldots k_n} = x_{m_{k_1 \ldots k_n}}.
\end{aligned}
\right.
\tag{2}
$$

The subscripts of the y in (2) are precisely all nonempty admissible sets of subscripts of the scheme S.

We shall say that f is a *superposition of the functions g* (with admissible sets of subscripts), if all functions g are defined on the unit cubes of the corresponding spaces and if all these functions (except perhaps the function g without subscripts) have values that satisfy $0 \leqslant g_{k_1 \ldots k_r} \leqslant 1$. Clearly, f is defined on the unit cube B: $0 \leqslant x_k \leqslant 1$, $k = 1, \cdots, s$.

We need the notion of a *type T of superpositions*. T is given by a scheme S and by an assignment of a class

$$
\Lambda = \Lambda_{k_1 \ldots k_r} = \Lambda_{p\omega}^s(M_0, \cdots, M_{p+1}; B) \subset \operatorname{Lip}_M 1
$$

for each admissible set of subscripts; the number s of variables must be $s = m_{k_1 \ldots k_r}$. Under these assumptions, T consists of all superpositions f that can be formed, according to (2), with functions $g_{k_1 \ldots k_r} \in \Lambda_{k_1 \ldots k_r}$. We see that T is a set of continuous functions f on the unit cube in R_s. With each type T, we associate the integer n—the *height* of T, and the number $M \geqslant 0$—the maximum of all M corresponding to the different classes Λ.

LEMMA 3. For each type T,

$$
H_{\epsilon_1}(T) \leqslant \sum H_{\epsilon}(\Lambda_{k_1 \ldots k_r}), \qquad \epsilon_1 = (M + 1)^n \epsilon,
\tag{3}
$$

where the sum is extended over all admissible sets of subscripts of the scheme S.

Proof. Let T_k, $k = 1, \cdots, m$, be the type of height $n - 1$ that is obtained from T in the following way: We remove the first row in (2) and fix the first subscript $k_1 = k$ in all remaining rows. In other words, the type T_k consists of all possible functions $f_k(x_1, \cdots, x_s) = g_k(y_{k1}, \cdots, y_{km_k})$, obtainable by (2), with functions $g_{k_1 \ldots k_r}$ in the classes $\Lambda_{k_1 \ldots k_r}$, and defined on the unit cube of the space R_s. For all $f \in T$,

$$
f = g(f_1, \cdots, f_m), \qquad g \in \Lambda, \qquad f_k \in T_k, \qquad k = 1, \cdots, m.
\tag{4}
$$

Let $\epsilon_2 = (M + 1)^{n-1} \epsilon$. Let U be the sets of some minimal ϵ-covering of Λ, $U^{(k)}$ the sets of minimal ϵ_2-coverings of T_k, $k = 1, \cdots, m$. If $f = g(f_1, \cdots, f_m)$, $f' = g'(f_1', \cdots, f_m')$, where g and g' belong to the same U, and where for each k, f_k and f_k' belong to the same $U^{(k)}$, then (see 3.7(5)),

$$| f(x_1, \cdots, x_s) - f'(x_1, \cdots, x_s) | \leqslant | g(f_1, \cdots, f_m) - g'(f_1, \cdots, f_m) |$$

$$+ | g'(f_1, \cdots, f_m) - g'(f_1', \cdots, f_m') | \leqslant \| g - g' \|$$

$$+ M \max_k \{ | f_k(x_1, \cdots, x_s) - f_k'(x_1, \cdots, x_s) | \} \leqslant 2\epsilon + M \cdot 2\epsilon_2 \leqslant 2\epsilon_1.$$

In this way we obtain an ϵ_1-covering of T, which consists of $N_\epsilon(\Lambda) \prod_{k=1}^{m} N_{\epsilon_2}(T_k)$ sets. Hence,

$$H_{\epsilon_1}(T) \leqslant H_\epsilon(\Lambda) + \sum_{k=1}^{m} H_{\epsilon_2}(T_k). \tag{5}$$

From this, (3) follows by iteration.   ∎

Returning to the proof of Theorem 3, we consider the Banach space $X = X_{p_0 \omega_0}^{s_0}$ (see Chapter 3, Sec. 7). A ball U of radius r in X is a translation of the ball with center the origin, and this ball is $\Lambda_{p_0 \omega_0}^{s_0} (r, \cdots, r; B)$. The entropy of U in the uniform norm can be estimated by 10.2(2). This gives $H_\epsilon(U) \geqslant K / \Delta_0(\beta \epsilon)$. Similarly, for each of the classes $\Lambda \in F$, $H_\epsilon(\Lambda) \leqslant K_1 / \Delta(\gamma \epsilon)$. Hence,

$$\frac{H_\epsilon(\Lambda)}{H_\epsilon(U)} \leqslant \text{const} \, \frac{\Delta_0(\beta \epsilon)}{\Delta(\gamma \epsilon)} \to 0 \qquad \text{as} \qquad \epsilon \to 0,$$

by (1). If the type T is formed of classes $\Lambda \in F$, then (3) implies that $H_\epsilon(T) / H_\epsilon(U) \to 0$. It follows that T does not contain any ball U of X.

Since the classes Λ are compact in the uniform topology (Theorem 9, Chapter 3), T is also compact and hence closed. Since convergence in the norm of X implies uniform convergence, $X \cap T$ is closed in the space X. It follows that $X \cap T$ is *nowhere dense* in X.

There are countably many schemes S; hence, there are countably many types formed with classes $\Lambda \in F$. For these T, the set $X \cap \cup T$ is of first category in X; hence, by Baire's theorem, $U \setminus \cup T$ is not empty for each ball U in X. In particular, there exists a function $f \in \Lambda_{p_0 \omega_0}^{s_0} \setminus \cup T$, which is not representable by superpositions of functions of L.   ∎

This proof gives a little more: In some sense "most" functions $f \in \Lambda_{p_0 \omega_0}^{s_0}$ do not have the representation in question.

5. Notes

1. It is not known whether the number $2s + 1$ of summands in the formulas 1(2) or 1(3) can be reduced. R. Doss [42] showed, however, that this is impossible for 1(3) if $s = 2$ and if all functions ϕ_{pq} are to be monotone increasing.

2.   On the other hand, it is known that the functions ϕ_p in the formula 1(3) cannot be very smooth. This follows from the following result, due to Vituškin [101, 102] and Henkin [56]. Let $\phi_q(x, y)$, $\psi_q(x, y)$, $q = 1, \cdots, N$ be continuous functions defined on the whole plane, and let the functions ϕ_q be continuously differentiable. Then there exists a polynomial $(x + my)^n$, with natural n, m, which cannot be represented in the form

$$\sum_{q=1}^{N} \psi_q(x, y) \, g_q(\phi_q(x, y))$$

for any selection of bounded measurable functions g_q . Unfortunately, we cannot prove this theorem here.

PROBLEMS

1.   Prove that the function $\psi(t)$ of p. 170 is a nonconvex modulus of continuity.

2.   Carry out the proof of Theorem 1 for an arbitrary s.

3.   Show that this proof still works if one assumes that λ_p are arbitrary numbers that satisfy $0 < \lambda_p < 1$, and for which a relation $r_1\lambda_1 + \cdots + r_s\lambda_s = 0$ with integers r_p is possible only if $r_1 = \cdots = r_s = 0$.

4.   Prove that Theorem 2 remains true if we add to L countably many functions, each belonging to a class $\text{Lip}_M 1$.

5.   Using Theorem 6 of Chapter 10, prove that not all analytic functions of s variables are superpositions of analytic functions of fewer variables.

6.   There are functions ϕ_q , $q = 0, \cdots, 4$, that satisfy Theorem 1 and belong to $\text{Lip } \alpha$ for each $0 < \alpha < 1$. [*Hint:* Change the construction that leads to the proof of Lemma 1 in the following way: At the kth step use the intervals I_{li}^q of rank $l = l(k) = k^2$.]

Bibliography

A. Books on Approximation

1. Achieser (Ahiezer), N. I., *Theory of Approximation*, New York: Ungar, 1956. (307 pp.) (Translated from the Russian.) [There exists an augmented 2nd edition, Moscow: Nauka, 1965. (407 pp.)]
2. Bernstein, S. N., *Collected Works* (Russian), Akad. Nauk SSSR, Moscow, vol. I, 1952 (581 pp.), vol. II, 1954. (627 pp.)
3. ——— *Leçons sur les propriétés extrémales et la meilleure approximation des fonctions analytiques d'une variable réelle*, Paris, 1926. (207 pp.) In: *l'Approximation*, New York: Chelsea, 1970.
4. Davis, Ph. J., *Interpolation and Approximation*, New York: Blaisdell, 1963. (393 pp.)
5. Gončarov, V. L., *Theory of Interpolation and of Approximation of Functions*, 2nd ed., Moscow: Gostehizdat, 1954. (327 pp.) (In Russian.)
6. Jackson, Dunham, *The Theory of Approximation*, Amer. Math. Soc. Colloquium Publications, vol. XI, New York, 1930. (178 pp.)
7. Korovkin, P. P., *Linear Operators and Approximation Theory*, Moscow: Fizmatgiz, 1959. There is also a not completely satisfactory English translation by Hindustan Publ. Corp., Delhi, 1960. (222 pp.)
8. Lorentz, G. G., *Bernstein Polynomials*, Toronto: Univ. of Toronto Press, 1953. (130 pp.)
9. Meinardus, G., *Approximation von Funktionen und ihre numerische Behandlung*, (Springer Tracts in Natural Philosophy, vol. 4), Berlin: Springer, 1964. (180 pp.)
10. Natanson, I. P., *Constructive Function Theory*, New York: Ungar, 1965. (Approx. 515 pp.) (Translated from the Russian.)
11. Rice, J. R., *The Approximation of Functions*, vol. 1, *Linear Theory*, Reading, Mass.: Addison-Wesley, 1964. (203 pp.)
12. Sard, A., *Linear Approximation*, Math. Surveys No. 9, Providence: Amer. Math. Soc., 1963. (544 pp.)
13. Smirnov, V. I., and N. I. Lebedev, *Constructive Theory of Functions of a Complex Variable*, Moscow-Leningrad: Nauka, 1964. (438 pp.) (In Russian.)
14. Stiefel, E. L., *An Introduction to Numerical Mathematics*, New York and London: Academic Press, 1963. (286 pp.)
15. Timan, A. F., *Theory of Approximation of Functions of a Real Variable*, New York: Macmillan, 1963. (631 pp.) (Translated from the Russian.)
16. de la Vallée-Poussin, C., *Leçons sur l'approximation des fonctions d'une variable réelle*, Paris, 1919. (151 pp.) In: *l'Approximation*, New York: Chelsea, 1970.

17. Vitushkin (Vituškin), A. G., *Theory of Transmission and Processing of Information*, New York: Pergamon, 1961. (206 pp.). (The Russian original had the title *Estimation of the Complexity of the Tabulation Problem*.)
18. Walsh, J. L., *Interpolation and Approximation by Rational Functions in the Complex Domain*, vol. XX, 2nd ed., Providence: Amer. Math. Soc. Coll. Publ., (1956). (398 pp.)
19. Walsh, J. L., *Approximation by Bounded Analytic Functions*, *Mémorial des Sci. Math.*, vol. 144, Paris: Gauthier-Villars, 1960. (66 pp.)
20. *On Approximation Theory, Proceedings of the Conference at Oberwolfach*, 1963, Basel: Birkhäuser, 1964. (261 pp.)
21. *Approximation of Functions, Proceedings of the Symposium, General Motors Research Laboratories, 1964*, Amsterdam: Elsevier, 1965. (220 pp.)

B. Some Other Books

22. Eggleston, H. G., *Convexity* (Cambridge Tracts No. 47), Cambridge: University Press, 1958.
23. Hardy, G. H., J. E. Littlewood, and G. Polya, *Inequalities*, 2nd ed., Cambridge: University Press, 1952.
24. Hardy, G. H., and W. W. Rogosinski, *Fourier series*, 2nd ed. (Cambridge Tracts No. 38), Cambridge: University Press, 1950.
25. Kneser, H., *Funktionentheorie*, Göttingen: Vandenhoeck and Ruprecht, 1958.
26. Natanson, I. P., *Theory of Functions of a Real Variable*, vol. 1, New York: Ungar, 1955.
27. Neumark, M. A., *Linear Differentialoperatoren*, Berlin: Akademie-Verlag, 1960.
28. Simmons, G. F., *Introduction to Topology and Modern Analysis*, New York: McGraw-Hill, 1963.
29. Whittaker, E. T., and G. N. Watson, *A Course in Modern Analysis*, 4th ed., Cambridge: University Press, 1952.

C. Papers in Journals

Abbreviations: American Mathematical Society: *TAMS* = Transactions, *BAMS* = Bulletin, and *PAMS* = Proceedings; *Izv.* = Izvestia Akad. Nauk SSSR, Ser. Mat.; *Dokl.* = Doklady Akad. Nauk SSSR; *Uspehi* = Uspehi Mat. Nauk.

30. Ahiezer (Achieser), N. I., and M. G. Kreĭn, "On the best approximation of periodic functions," *Dokl.*, 15 (1937), 107-112.
31. Al'per, S. Ja., "Asymptotic values of best approximation of analytic functions in a complex domain," *Uspehi*, 14, No. 1 (85) (1959), 131-134.
32. Babenko, K. I., "On the best approximation of a class of analytic functions," *Izv.* 22 (1958), 631-640.
33. Bari N. K., and S. B. Stečkin, "Best approximation and differential properties of two conjugate functions," *Trudy Mosk. Mat. Obšč.*, 5 (1956), 483-522.

34. Berman, D. L., "On a class of linear operators," *Dokl.*, **85** (1952), 13-16.
35. Bernstein, S. N., "Sur l'ordre de la meilleure approximation des fonctions continues par des polynomes de degré donné," *Mémoires publiés par la classe des sci. Acad. de Belgique* (2) **4** (1912), 1-103 (= [2, vol. I, 11-104]).
36. Birkhoff, G. D., "On the asymptotic character of the solutions of certain linear differential equations," *TAMS*, **9** (1908), 219-231.
37. Brudnyĭ, Ju. A., "Generalization of a theorem of A. F. Timan," *Dokl.*, **148** (1963), 1237-40.
38. Cheney, E. W., "Approximation by generalized rational functions," in [21], 101-110.
39. Clements, G. F., "Entropies of sets of functions of bounded variation," *Canadian J. Math.*, **15** (1963), 422-432.
40. Curtis, P. C., "*n*-parameter families and best approximation," *Pacific J. Math.*, **9** (1959), 1013-1027.
41. Dolženko, E. P., "Estimates of derivatives of rational functions," *Izv.*, **27** (1963), 9-28.
42. Doss, R., "On the representation of continuous functions of two variables by means of addition and continuous functions of one variable," *Colloquium Math.*, **10** (1963), 249-259.
43. Dzjadyk, V. K., "Constructive characterization of functions satisfying a condition Lip α $(0 < \alpha < 1)$ on a finite interval of the real axis," *Izv.*, **20** (1956), 623-642.
44. ——— "A further strengthening of Jackson's theorem on the approximation of continuous functions by ordinary polynomials," *Dokl.*, **121** (1958), 403-406.
45. Erdös, P., "Extremal properties of derivatives of polynomials," *Ann. Math.*, (2) **41** (1940), 310-313.
46. Erohin, V. D., "On the asymptotic behavior of the ϵ-entropy of analytic functions, *Dokl.*, **120** (1958), 949-952.
47. ———, "On the best approximation of analytic functions by rational functions with free poles," *Dokl.*, **128** (1959), 29-32.
48. Faber, G., "Über die interpolatorische Darstellung stetiger Funktionen," *Jahresber. DMV*, **23** (1914), 192-210.
49. Favard, J., "Sur les meilleures procédés d'approximation de certaines classes des fonctions par des polynômes trigonométriques," *Bull. Sci. Math.*, **61** (1937), 209-224, 243-256.
50. ——— "Sur la saturation des procédés de sommation," *J. Math. Pures Appl.*, (9) **36** (1957), 359-372.
51. Freud, G., "Über die Approximation reeller stetiger Funktionen durch gewöhnliche Polynome," *Math. Ann.*, **137** (1959), 17-25.
52. Gelfand, A. O., "On uniform approximation by polynomials with integral rational coefficients," *Uspehi*, **10**, No. 1 (63) (1955), 41-65.
53. Gohberg, I. C., and M. G. Kreĭn, "Fundamental aspects of defect numbers, root numbers, and indexes of linear operators," *Uspehi*, **12**, No. 2 (74) (1957), 43-118.
54. Gončar, A. A., "Inverse theorems of best approximation by rational functions," *Izv.*, **25** (1961), 347-356.

55. Haar, A., "Die Minkowskische Geometrie und die Annäherung stetiger Funktionen," *Math. Ann.*, **78** (1918), 294-311.

56. Henkin, G. M., "Linear superpositions of continuously differentiable functions," *Dokl.*, **157** (1964), 288-290.

57. Hewitt, E., and H. S. Zuckerman, "Approximation by polynomials with integral coefficients, a reformulation of the Stone-Weierstrass theorem," *Duke Math. J.*, **26** (1959), 305-324.

58. Jackson, D., "On the approximation by trigonometric sums and polynomials," *TAMS*, **13** (1912), 491-515.

59. Jurkat, W. B., and G. G. Lorentz, "Uniform approximation by polynomials with positive coefficients," *Duke Math. J.*, **28** (1961), 463-474.

60. Kolmogorov, A. N., "Über die beste Annäherung von Funktionen einer gegebenen Funktionenklasse," *Ann. of Math.*, (2) **37** (1936), 107-111.

61. ———— "A remark on the polynomials of Chebyshev, deviating the least from a given function," *Uspehi*, **3**, No. 1 (23) (1948), 216-221.

62. ———— "Asymptotic characteristics of some completely bounded metric spaces," *Dokl.*, **108** (1956), 585-589.

63. ————, "On the representation of continuous functions of several variables by superpositions of continuous functions of one variable and addition," *Dokl.*, **114** (1957), 953-956.

64. ————, "On the linear dimension of linear topological spaces," *Dokl.*, **120** (1958), 239-241.

65. Kolmogorov, A. N., and V. M. Tihomirov, "ε-entropy and ε-capacity of sets in function spaces," *Uspehi*, **14**, No. 2 (86) (1959), 3-86.

66. Korneĭčuk, N. P., "The best approximation of continuous functions," *Izv.*, **27** (1963), 29-44.

67. ————, "The best uniform approximation of differentiable functions," *Dokl.*, **141** (1961), 304-307.

68. ————, "On the existence of a linear polynomial operator which gives best approximation on a class of functions," *Dokl.*, **143** (1962), 25-27.

69. ————, "The exact constant in the theorem of D. Jackson on the best uniform approximation of continuous periodic functions," *Dokl.*, **145** (1962), 514-515.

70. ————, "The exact value of the best approximation and of widths of some classes of functions," *Dokl.*, **150** (1963), 1218-1220.

71. Kreĭn, M. G., D. Milman, and M. Rutman, "On a property of a basis in a Banach space," *Zapiski Harkov Mat. Obšč.*, (4) **16** (1940), 106-110.

72. Lax, P., "Proof of a conjecture of P. Erdös," *BAMS*, **50** (1944), 509-513.

73. Lorentz, G. G., "Lower bounds for the degree of approximation," *TAMS*, **97** (1960), 25-34.

74. ————, "Metric entropy, widths, and superpositions of functions," *Amer. Math. Monthly*, **69** (1962), 469-485.

75. ————, "The degree of approximation by polynomials with positive coefficients," *Math. Ann.*, **151** (1963), 239-251.

76. ————, "Inequalities and the saturation classes of Bernstein polynomials," in [20], 200-207.

77. ————, "Entropy and approximation," *BAMS*, to be published.

78. Mairhuber, J., "On Haar's theorem concerning Chebyshev approximation problems having unique solutions," *PAMS*, **7** (1956), 609-615.

79. Markov, A. A., "On a problem of D. I. Mendeleev," *St. Petersburg, Izv. Akad. Nauk*, **62** (1889), 1-24.

80. Mitjagin, B. S., "The approximative dimension and bases in nuclear spaces," *Uspehi*, **16**, No. 4 (100) (1961), 63-132.

81. ———, "Approximation of functions in L^p and C spaces on the torus," *Mat. Sbornik (N.S.)*, **58** (100) (1962), 397-414.

82. Newman, D. J., "Rational approximation to $|x|$," *Michigan Math. J.*, **11** (1964), 11-14.

83. Newman, D. J., and H. S. Shapiro, "Jackson's theorem in higher dimensions," in [20], 208-219.

84. Nikol'skiĭ, S. M., "On the best approximation of functions satisfying a Lipschitz's condition by polynomials," *Izv.*, **10** (1946), 295-322.

85. Phelps, R. R., "Uniqueness of Hahn-Banach extensions and unique best approximation," *TAMS*, **95** (1960), 238-255.

86. Rivlin, T. J., and H. S. Shapiro, "A unified approach to certain problems of approximation and minimization," *J. Soc. Indust. and Appl. Math.*, **9** (1961), 670-699.

87. Scheick, J. T., "Polynomial approximation of functions analytic in a disc," *PAMS*, **17** (1966), 1238-1243.

88. Schoenberg, I. J., "Spline interpolation and best quadrature formulae," *BAMS*, **70** (1964), 143-148.

89. Sprecher, D., Ph.D. Dissertation, University of Maryland, 1963.

90. Stečkin, S. B., "On the order of the best approximation of continuous functions," *Izv.*, **15** (1951), 219-242.

91. Stone, M. H., "The generalized Weierstrass approximation theorem," *Math. Magazine*, **21** (1948), 167-184, 237-254.

92. Sunouchi, G., and C. Watari, "On determination of the class of saturation in the theory of approximation of functions," *Tôhoku Math. J.*, **11** (1959), 480-488.

93. Tihomirov, V. M., "Widths of sets in functional spaces and the theory of best approximations," *Uspehi*, **15**, No. 3 (93) (1960), 81-120.

94. ———, "On the ε-entropy of certain classes of periodic functions," *Uspehi*, **17**, No. 6, (108) (1962), 163-169.

95. ———, "Some problems in approximation theory," *Dokl.*, **160** (1965), 774-777.

96. Timan, A. F., "A strengthening of Jackson's theorem on the best approximation of continuous functions by polynomials on a finite interval of the real axis," *Dokl.*, **78** (1951), 17-20.

97. ———, "A geometric problem in the theory of approximation," *Dokl.*, **140** (1961), 307-310.

98. Tucker, A. W., "Some topological properties of disc and sphere," *Proc. First Canadian Math. Congress*, Montreal (1945), 285-309.

99. Tureckiĭ, A. H., "On saturation classes for certain methods of summation of Fourier series of continuous periodic functions," *Uspehi*, **15**, No. 6 (96) (1960), 149-156.

100. Vituškin (Vitushkin), A. G., "On the 13th problem of Hilbert," *Dokl.*, **95** (1954), 701-704.
101. ———, "Some properties of linear superpositions of smooth functions," *Dokl.*, **156** (1964), 1003-1006.
102. ———, "Proof of existence of analytic functions of several variables, not representable by linear superpositions of continuously differentiable functions of fewer variables," *Dokl.*, **156** (1964), 1258-1261.
103. Weierstrass, K., "Über die analytische Darstellbarkeit sogennanter willkürlicher Funktionen reeller Argumente," *Sitzungsberichte der Acad. Berlin* (1885), 633-639, 789-805.
104. Whittlesey, E. F., "Fixed points and antipodal points," *Amer. Math. Monthly*, **70** (1963), 807-821.
105. Žuhovickiĭ, S. I., "On approximation of real functions in the sense of P. L. Chebyshev," *Uspehi*, **11**, No. 2 (68) (1956), 125-159.
106. Zygmund, A., "Smooth functions," *Duke Math. J.*, **12** (1945), 47-76.

Index

Index

CHELSEA

SCIENTIFIC

BOOKS

NUMERICAL MATHEMATICS

BERNSTEIN, Serge: Leçons sur les Propriétés Extrémales et la Meilleure Approximation des Fonctions Analytiques d'une Variable Réelle; & VALLÉE POUSSIN, C. de la: Leçons sur L'Approximation des Fonctions d'une Variable Réelle. 2 vols in 1. (French) 363 pp. 6 x 9. ISBN -0198-5, ∞G

BOOLE, George: Treatise on the Calculus of Finite Differences, 5th ed. 341 pp. 5⅜ x 8. ISBN -1121-2, ∞G

CHENEY, Ward: Introduction to Approximation Theory, 2nd ed. x + 260 pp. 6 x 9. ISBN -0317-1, ∞G

JORDAN, Charles: The Calculus of Finite Differences, xxi + 655 pp. 5⅜ x 8. ISBN -0033-4, ∞G

LORENTZ, George G.: Approximation of Functions, ix + 184 pp. 6 x 9. ISBN -0322-8, ∞G

MILNE-THOMSON, L.M.: The Calculus of Finite Differences. 2nd (unaltered) ed. xix + 558 pp. 5⅜ x 8, ISBN -0308-2, ∞G

NOERLUND, Niels H.: Differenzenrechnung. (Germ.) ix + 551 pp. 5⅜ x 8. ISBN -0100-4, ∞'G

RUNGE, Carl: Graphical Methods. Included in SIERPINSKI, W.: Congruence of Sets. (General Math.)

TRAUB, J. F.: Iterative Methods for the Solution of Equations, 2nd (unaltered) ed. xv + 310 pp. 6 x 9. ISBN -0312-0, ∞G

PROBABILITY AND STATISTICS

BERTRAND, Joseph: Calcul des Probabilites. 3rd ed. (French) 57 + 322 pp. 5⅜ x 8. ISBN -0262-0, ∞G

CONDORCET, M. J.: Essai sur L'Application de L'Analyse aux Probabilites. (French) 191 + 304 pp. 6 x 9. ISBN -0252-3, ∞G

DE MOIVRE, Abraham: The Doctrine of Chances, 3rd ed. xii + 368 pp. 6 x 9. ISBN -0200-0, ∞G

— — —A Treatise on the Annuities of Lives. included in: DE MOIVRE: The Doctrine of Chances, 3rd ed.

GNEDENKO, Boris V.: The Theory of Probability and the Elements of Statistics (with answers to the exercises), 5th ed., 527 pp. 6 x 9. ISBN -2320-1, ∞

GRENANDER, Ulf & ROSENBLATT, Murray: Statistical Analysis of Stationary Time Series, 2nd (corr.) ed. 308 pp. 6 x 9. ISBN -0320-1, ∞G

GRENANDER, Ulf & SZEGO, Gabor: Toeplitz Forms and Their Applications, 2nd (corr.) ed. x + 245 pp. 5⅜ x 8. ISBN -0321-X, ∞G

KOLMOGOROV, Andrei N.: Foundations of the Theory of Probability. viii + 84 pp. 6 x 9. ISBN -0023-7, ∞G

MONTMORT, R. de: Essay sur les Jeux de Hazard, (French) 458 pp. + plates. 6½ x 10. ISBN -0307-4, ∞G

TODHUNTER, Isaac: History of the Mathematical Theory of Probability. xvi + 624 pp. 5⅜ x 8. ISBN -0057-1, ∞G

VENN, John: The Logic of Chance, 4th ed. xxix + 508 pp. 5⅜ x 8. ISBN -0173-X, ∞G

WALD, Abraham: Statistical Decision Functions. ix + 179 pp. 5⅜ x 8. ISBN -0243-4, ∞G